教育部高职高专计算机类专业教学指导委员会规划教材

# 网络设备与集成

刘福新　陈小中　编著

中国铁道出版社
CHINA RAILWAY PUBLISHING HOUSE

## 内 容 简 介

本书立足"认识网络、感知网络、管理网络"的指导思想，依据典型校园网络建设的实际工作内容和操作流程，利用思科公司 Packet Tracer 模拟软件，结合作者十多年网络教学和工程项目经验，在网络设备原厂研发工程师的帮助下，历时 2 年完成编著工作。本书共 8 章，内容涉及网络规划设计、VLAN、OSPF、ACL、DHCP、DNS、VPN、IPv6 和常用网络管理工具等，将网络规划、建设和管理工作中经常使用的知识和技能浓缩在 19 个项目中，构建"教、学、做"一体化学习环境，让每个学生都可以掌控一个校园网，分享网络工程师的感受，体验网络管理员的工作。

本书适合作为高等职业院校计算机网络及相关专业的教材，也可作为一线网络工程技术人员的参考书。

**图书在版编目（CIP）数据**

网络设备与集成 / 刘福新，陈小中编著. -- 北京：
中国铁道出版社，2011.7
教育部高职高专计算机专业教学指导委员会规划教材
ISBN 978-7-113-13066-4

Ⅰ.①网… Ⅱ.①刘… ②陈… Ⅲ.①校园网—高等职业教材 Ⅳ.①TP393.18

中国版本图书馆 CIP 数据核字（2011）第 145140 号

| | | | |
|---|---|---|---|
| 书 名：| 网络设备与集成 | | |
| 作 者：| 刘福新 陈小中 编著 | | |
| 策 划：| 翟玉峰 巨 凤 | 读者热线： | 400-668-0820 |
| 责任编辑：| 翟玉峰 | 特邀编辑： | 李红玉 |
| 编辑助理：| 何 佳 | 封面制作： | 白 雪 |
| 封面设计：| 付 巍 | 责任印制： | 李 佳 |

出版发行：中国铁道出版社（100054，北京市宣武区右安门西街 8 号）
网　址：http://www.tdpress.com　http://www.edusources.net
印　刷：北京市兴顺印刷厂
版　次：2011 年 8 月第 1 版　　2011 年 8 月第 1 次印刷
开　本：787mm×1092mm　1/16　印张：14.25　字数：339 千
印　数：1～3 000 册
书　号：ISBN 978-7-113-13066-4
定　价：25.00 元

**版权所有　侵权必究**

凡购买铁道版图书，如有印制质量问题，请与本社教材研究开发中心批销部联系调换。电话：(010) 63550836
打击盗版举报电话：(010) 63549504

**教育部高职高专计算机类专业教学指导委员会规划教材**

编审委员会

主　任：温　涛

副主任：孙　湧　严晓舟

编　委：(按姓氏笔画排序)

| | | | | |
|---|---|---|---|---|
| 丁桂芝 | 王　勇 | 王公儒 | 石　硕 | 史宝会 |
| 刘甫迎 | 刘晓川 | 刘海军 | 刘福新 | 安志远 |
| 许洪军 | 杨洪雪 | 杨俊清 | 吴建宁 | 邱钦伦 |
| 邹　翔 | 宋汉珍 | 张晓云 | 陈　晴 | 赵凤芝 |
| 胡昌杰 | 秦绪好 | 徐　红 | 褚建立 | 翟玉峰 |

# 序

PREFACE

《国家中长期教育改革和发展规划纲要(2010—2020年)》文件指出,职业教育要面向人人、面向社会,着力培养学生的职业道德、职业技能和就业创业能力。到2020年,形成适应经济发展方式转变和产业结构调整要求、体现终身教育理念、中等和高等职业教育协调发展的现代职业教育体系,满足人民群众接受职业教育的需求,满足经济社会对高素质劳动者和技能型人才的需要。

高等职业教育肩负着培养生产、建设、服务和管理第一线高素质技能型专门人才的重要使命,在对经济发展的贡献方面具有独特作用。十多年来,我国高等职业教育规模迅速扩大,为实现高等教育大众化发挥了积极作用。同时,高等职业教育也主动适应社会需求,坚持以服务为宗旨,以就业为导向,走产学研结合发展的道路,切实把改革与发展的重点放到加强内涵建设和提高教育质量上来,更好地为我国全面建设小康社会和构建社会主义和谐社会,建设人力资源强国做出贡献。自1998年以来,我国高职院校培养的毕业生已超过1 300万人,为经济领域内的各行各业生产和工作第一线培养了大批高素质技能型专门人才。目前,全国高等职业院校共有1 200余所,年招生规模达到310万人,在校生达到900万人;高等职业院校招生规模占到了普通高等院校招生规模的一半,已成为我国高等教育的"半壁江山"。

《关于全面提高高等职业教育教学质量的若干意见》教高[2006]16号文件指出,课程建设与改革是提高教学质量的核心,也是教学改革的重点和难点。高等职业院校要积极与行业企业合作开发课程,根据技术领域和职业岗位(群)的任职要求,参照相关的职业资格标准,改革课程体系和教学内容。建立突出职业能力培养的课程标准,规范课程教学的基本要求,提高课程教学质量。文件中还指出,与行业企业共同开发紧密结合生产实际的实训教材,并确保优质教材进课堂。重视优质教学资源和网络信息资源的利用,把现代信息技术作为提高教学质量的重要手段,不断推进教学资源的共建共享,提高优质教学资源的使用效率,扩大受益面。

为落实教高[2006]16号文件精神,教育部高等学校高职高专计算机类专业教学指导委员会(简称"计算机教指委")于2009年11月19日在陕西西安召开"高职高专计算机网络专业教学改革研讨会",就高职高专计算机网络专业的专业建设、教学模式、课程设置、教材建设等内容进行了研讨,确定了计算机网络技术专业建设的三个方向:即计算机网络工程与管理、计算机网络安全和网站规划与开发。2010年计算机教指委承办的全国职业院校技能大赛高职组的"计算机网络组建与安全维护"竞赛,对未来高等职业教育计算机网络专业的改革和发展也起到了重要的促进作用。

中国铁道出版社为配合落实《国家中长期教育改革和发展规划纲要(2010—2020年)》,贯彻全国高等职业教育改革与发展工作会议精神,与计算机教指委合作,组织高职院校一线教师及行业企业共同开发了这套计算机网络技术专业教材。本套教材以课程建设为核心,以教育部计算机网络大赛为契机,本着以服务为宗旨,以就业为导向,积极围绕职业岗位人才需求的总目标和职业能力需求,根据不同课程在课程体系中的地位及作用,根据不同工作过程,将课程内容、教学方法和手段与课程教学环境相融合,形成了以工作过程对知识的基本要求为主体的围绕问题中心的教材和以基础能力训练为核心的围绕基础训练任务的教材、以岗位综合能力训练为核心、以任务为中心的教材等多种教材编写形式。

网络信息的发展，给社会的发展提供了动力，高职高专教育要随时跟上社会的发展，抓住机遇，培养适合我国经济发展需求、能力符合企业要求的高素质技能型人才，为我国高职高专教育的发展添砖加瓦。希望通过本套教材的出版，为推广高职高专教学改革，实现优秀教学资源共享，提高高职高专教学质量，向社会输送高素质技能型人才做出更大贡献。

温 涛

2011 年 1 月

# 前言

## FOREWORD

不知不觉与网络打交道已经十多年了，经历了学校校园网的规划、建设和管理，参与了许多园区网建设方案的研讨，讲授过网络专业学生的专业课程，指导过网络专业学生的毕业设计，如能把这些年的心得体会和经验总结写出来与各位分享，让读者更加便捷高效地学习网络的相关知识，岂不是一件好事？

作者曾利用全套网络实训装置进行实训教学，但由于设备台/套数有限，一个学生无法完全掌控一个网络，更不能架构一个接近真实情况的网络。学生所参与的实训大多是单项功能局部实施，无法体会网络工程师在规划和建设过程中的艰辛，也不能感受网络管理员在运行和维护工作中的无奈，更多的是验证性实验。很多学校在实践环境建设方面，不惜花巨资购置高档网络实训设备，但从教学效果来看，优越的环境并没有促进学习效果的提升。作者也尝试过利用网络设备原厂认证工程师培训的内容和方式进行教学，并借助 Packet Tracer 模拟软件进行实训体验，同时配合真实设备配置调试，但由于实训项目的组织缺乏连续性和全局性，学生在学完后只知道虚拟局域网和 OSPF 动态路由概念，少量设备互连时能进行验证配置，但一遇到实际工程问题时就无从下手，永远没有部署园区网络的体验。

因此，高职院校急需开发一本内容简洁、项目真实、操作性强、实用性好、能较好反映实际工作内容和操作规范的"网络设备与集成"教材。为此，作者与网络设备原厂研发工程师进行了为期两年的框架体系探讨，分析网络设备与集成工作中需要掌握的主要知识和技能，结合典型校园网工程精心设计教学项目，基于工作过程对教材进行设计。

本书以校园网设计、实施和运维工作的实际需求为导向进行项目任务设计，学生主要利用思科公司 Packet Tracer 模拟软件进行训练，如有条件可利用硬件平台进行验证训练效果更好。作者力求用简洁明了的语言文字，图文并茂的版面风格，通俗易懂的操作指导，让读者在快速轻松的阅读过程中获取核心知识，在反复训练的操作过程中培养职业技能。本书共 8 章，每章设计了多个项目，每个项目基本按照"项目描述—知识准备—项目实施—工程化操作"的体系编写，在重点分析"项目实施"后，通过"工程化操作"强调工程实施的方法与规范。在内容安排上除了体现实际项目的工作内容和操作规范外，更加关注职业院校学生的学习心态。

本书第 1 章介绍校园网结构与应用服务，让学生利用 Packet Tracer 动手搭建一个属于自己的校园网络，按照脚本手工配置网络设备使网络动起来。学生在实践中去理解通信介质、通信协议、网络设备、接口类型、扩展模块、设备间、数据中心和拓扑结构等概念，去体验网络工程师的工作内容及方法，去感受校园网的应用服务。当出现路由器没有接口连接光缆、打错一个字母致使网络连接失败、交换机不在实施现场等情况时，学生应马上问老师学友、读在线帮助、搜网上资源，想尽一切办法去解决问题。第 2 章介绍虚拟局域网，在园区网络范围部署各种 VLAN，让学生体验接入 VLAN、汇聚 VLAN、管理 VLAN 的概念和配置方法，通过新增楼宇培养网络可扩展性设计理念。第 3 章介绍内网优化与管理，包括 OSPF 路由设计、访问限制列表和安全机制等，重点分析 OSPF 路由的部署，典型 ACL 与端口安全等。第 4 章介绍校园网应用服务，训练学生在园区网络中部署 DHCP 和 DNS 服务的能力。第 5 章介绍校园网出口设计，包括 VPN 和 CBAC 的主要技术与部署方案。第 6 章介绍校园网运维方式,包括远程登录、TFTP、SNMP、RSPAN、AAA 认证等内容。第 7 章介绍 IPv6 的编址与 IPv4 的过渡方法。第 8 章主要结合第 1～7 章的内容,根据工程项目中的不同设备品牌需求、以不同 IOS 软件版本和模拟软件局限性,设计了 4 个项目供学生训练,以提醒读者要理论联系实际、跟踪技术发展动向,利用原厂网站技术文献资源,学习解决工程实际问题能力。全书反复训练学生在校园网系统集成过程中需要用到的关键技能和职业素养,书中提供的所有配置文件和配置脚本全部在 Packet Tracer 5.3 中调试通过。

使用本书前建议读者先学习计算机网络的基础知识,建议按 72 学时组织教学,教师可根据实际情况进行微调。

本书由常州工程职业技术学院刘福新负责教材编写方案设计和项目策划,陈小中老师负责行文和项目调试。本书适合作为高等职业院校计算机网络及相关专业的教材,也可作为一线网络工程技术人员的参考书。

感谢网络设备原厂工程师赵广、余福生对本书编写工作付出的辛勤劳动,感谢中国铁道出版社的鼎力相助,尤其令人感动的是,在书稿提交比预定时间推迟的情况下,他们为保证教材的出版进度及对教材质量的把关所付出的劳动。最后,还要对本书在编写过程中所参考的国内外文献的诸多作者表示感谢。

由于编者水平所限,书中难免有疏漏之处,恳请读者及专家不吝赐教。

编　者
2011 年 6 月

# 目 录

## 第1章 校园网结构与应用服务 ... 1
- 项目1 建设校园网基础 ... 1
- 项目2 配置校园网 ... 10
- 本章训练内容 ... 43

## 第2章 使用VLAN部署校园网 ... 44
- 项目3 部署接入VLAN ... 44
- 项目4 部署汇聚VLAN ... 56
- 项目5 部署管理VLAN ... 73
- 本章训练内容 ... 78

## 第3章 校园网内网优化与管理 ... 79
- 项目6 优化OSPF路由 ... 79
- 项目7 访问控制管理 ... 113
- 项目8 内网安全 ... 121
- 本章训练内容 ... 128

## 第4章 校园网应用服务 ... 129
- 项目9 部署DHCP系统 ... 129
- 项目10 部署DNS系统 ... 143

## 第5章 校园网出口设计 ... 149
- 项目11 部署VPN ... 149
- 项目12 部署防火墙 ... 163

## 第6章 校园网运行维护 ... 168
- 项目13 使用管理工具 ... 168
- 项目14 AAA部署 ... 180

## 第7章 IPv6部署 ... 184
- 项目15 IPv6实验网 ... 184

## 第8章 综合训练 ... 193
- 项目16 物理架构与扩展 ... 193
- 项目17 路由协议分析 ... 196
- 项目18 新增应用服务 ... 198
- 项目19 H3C解决方案 ... 200

附录 项目训练图 ... 201
参考文献 ... 216

# 第 1 章 校园网结构与应用服务

校园网是一种典型的园区网，随着校园信息化建设的发展，校园网中部门、人员数量的增多，各种应用需求对校园网建设提出了很多挑战，包含教务系统、邮件系统、办公系统、网站、域名系统等，尤为重要的是核心业务、数据安全的保证问题。

通过合理设计、规划校园网可以建成快速、高效的校园网。通常需要对校园网的物理分布，以及部门与用户数进行分析计算，合理安排各个楼宇机房与设备间布局，部署各个部门网段及数据中心等。

通过本章中项目的实践，读者可以学会校园网内楼宇与设备间设备的部署，通过配置脚本对设备进行配置，体验校园网的各种服务，重点在于体验校园网。

本章需要完成的项目有：

项目 1——建设校园网基础；

项目 2——配置校园网。

## 项目 1　建设校园网基础

### 项目描述

随着校园信息化建设的发展，校园网的用户规模已超万人，校园网分布见图 1-1。为了有效地建设校园网，需要对校园网的物理分布进行分析，例如，如何设计楼宇设备间？采用何种链路连接校园网？校园网中常用的是哪些设备？此类问题将在本项目中进行分析、解决。

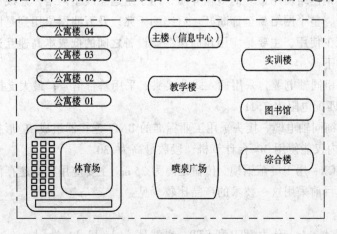

图 1-1　校园网分布图

**知识准备**

1）园区网络传输介质简介

传输介质按照有线和无线可以进行如下分类。

① 有线传输介质：双绞线、同轴电缆、光纤。

② 无线传输方式：无线电波、微波、红外线、激光、蓝牙。

衡量传输介质性能的主要技术指标有：传输距离、传输带宽、衰减、抗干扰能力等。

无线传输方式主要用在有线网络无法延伸到或者架有线网络耗资比较大的区域，而在园区网中主要就是使用有线传输介质，因此我们着重熟悉有线传输这块，有线传输的主要的 IEEE 标准如表 1-1 所示。

表 1-1  有线传输的传输标准

| 标　　准 | MAC 子层规范 | 最 大 长 度/m | 类　　　　型 | 对数 |
| --- | --- | --- | --- | --- |
| 10Base5 | 802.3 | 500 | 50Ω thick coaxial cable | |
| 10Base2 | 802.3 | 185 | 50Ω thin coaxial cable | |
| 10Base-T | 802.3 | 100 | Category3,4,or5 UTP | 2 |
| 100Base-FL | 802.3 | 2000 | Fiber | 1 |
| 100Base-TX | 802.3u | 100 | Category5 UTP | 2 |
| 100Base-T4 | 802.3u | 100 | Category3 UTP | 4 |
| 100Base-T2 | 802.3u | 100 | Category3,4,or5 UTP | 2 |
| 10Base-FX | 802.3u | 400/2000 | Multimode fiber | 1 |
| 100Base-FX | 802.3u | 10 000 | single-mode fiber | 1 |
| 1000Base-FX | 802.3z | 220～550 | Multimode fiber | 1 |
| 1000Base-LX | 802.3z | 3000 | Single-mode fiber or Multimode fiber | 1 |
| 1000Base-CX | 802.3z | 25 | Shielded copper | 2 |
| 1000Base-T | 802.3ab | 100 | Category 5 UTP | 2 |

2）同轴电缆

同轴电缆是早期网络使用最多的传输介质，用于组建总线型网络，总线型网络最大的缺点是：只要一点断开，整个网络就全部断开。后续的千兆以及万兆网络用到是同轴铜缆，但是在实际环境中仍然极少使用，主要是应用在高速存储设备之间的低成本高速互连。同轴电缆的传输标准以及特性如下。

① 10Base5：粗同轴电缆，采用插入式分接头；采用基带信号；最大支持段长为 500 m，最多段数为 100；匹配电阻为 75Ω。

② 10Base2：细同轴电缆，接头采用工业标准的 BNC 连接器组成 T 形插座；使用范围只有 200 m，每一段内仅能使用 30 台计算机，段数最高为 30。

③ 1000Base-CX：用于屏蔽铜缆，传输距离为 25 m。主要应用在高速存储设备之间的低成本高速互连，不过目前采用这一技术的产品比较少见。

3）双绞线

双绞线（twisted-pair）分为 UTP 和 STP，也就是 unshielded twisted-pair（非屏蔽双绞线）

和 shielded twisted-pair（屏蔽双绞线）。双绞线内部总共有八芯四对，分别为：橙白、橙、绿白、蓝、蓝白、绿、棕白、棕，双绞线基本都是采用 RJ-45 连接器，由 568B 和 568A 两种线序标准，具体的传输标准如下。

① 10Base-T：双绞线电缆，一般使用 RJ-45 连接器；最大有效传输距离是距集线器 100 m，即使是高质量的五类双绞线也只能达到 150 m。

② 100Base-TX：使用五类以上双绞线，最大传输距离为 100 m。

③ 1000Base-T：定义在传统的五类双绞线上，传输距离为 100 m，应用于高速服务器和工作站的网络接入，也可作为建筑物内的千兆骨干连接。

4）光纤

光纤是平时进行项目实施时采用最多的传输介质，也是最容易搞混淆的，尤其是光纤接口类型、工作模式、传输距离等，接下来将进行详细的讲解。

光纤的类型，可以分为单模和多模。

① 单模：当光纤的几何尺寸可以与光波长相比拟时，即纤芯的几何尺寸与光信号波长相差不大时，一般为 5～10μm，只允许以一种模式在其中传播信号。单模光纤具有极宽的带宽，特别适用于大容量、长距离的光纤通信。

② 多模：多模光纤纤芯的几何尺寸远大于光波波长，一般为 50μm、62.5μm，光信号是以多种模式进行传播的。多模光纤仅用于较小容量、短距离的光纤通信。

 注意

单模光可在多模光纤中传输，但多模光不能在单模光纤中传输。

光纤尾纤以及光模块的接口类型常用的主要有：SC、LC、ST、FC、MT-RJ。

光纤的传输标准主要为：百兆、千兆以及万兆。

光纤传输有长波和短波之分，主要的差别如下：

- 长波：长波的光信号波长在 1310～1550 nm 之间，因具有衰减低、带宽宽等优点，适用于长距离、大容量的光纤传输。
- 短波：短波的光信号波长在 600～900 nm 之间。适用于短距离、小容量的光纤传输。

（1）百兆光纤传输标准

百兆传输标准为 IEEE 802.3u，百兆光纤中的 S 代表 SINGLE，传输标准如表 1-2 所示。

表 1-2　百兆光纤传输标准

| 光纤协议 | 光纤标准 | 光纤尺寸/μm | 传输波长/nm | 传输距离/km |
| --- | --- | --- | --- | --- |
| 100Base-FX | 多模 | 62.5/125 | 1310 | 2 |
| 100Base-FX-S | 单模 | 9/125 | 1310 | 30 |

（2）千兆光纤传输标准

IEEE 802.3z 分别定义了三种千兆传输标准：1000Base-LX、1000Base-SX、1000Base-CX。千兆光纤中的 S 代表 Short，即短波，只可接多模光纤。千兆光纤中的 L 代表 Long，即长波，可接单模、多模光纤。千兆传输标准如表 1-3 所示。

表1-3 千兆光纤传输标准

| 光纤协议 | 光纤标准 | 光纤尺寸/μm | 传输波长/nm | 传输距离/km |
| --- | --- | --- | --- | --- |
| 1000Base-SX | 多模 | 62.5/125 | 850 | 0.22 |
|  |  | 50/125 |  | 0.5 |
| 1000Base-LX | 多模 | 62.5/125 | 1310 | 0.55 |
|  |  | 50/125 |  |  |
|  | 单模 | 9/125 | 1310 | 10 |
| 1000Base-LH | 单模 | 9/125 | 1310 | >70 |
| 1000Base-ZX | 单模 | 9/125 | 1550 | 70~100 |

工作在普通单模光纤链路上，最大传输距离达70以上，必须与单模光纤一起使用，这种光纤通常用在长距离电信应用中。不能与多模光纤配合使用，因此，在那些经常使用多模光纤的应用环境（如楼宇的主干、水平布线）中，不能使用1000BASE-ZX。

 注意

当使用短距离的单模光纤时，在链路中应该插入一个线上光衰减器以免光接收机过载。

（3）万兆光纤传输标准

IEEE 802.3ae 是10GE的标准，802.3ae 目前支持9μm单模、50μm多模和62.5μm多模三种光纤，增加了新的编码方式64B/66B（传统千兆以太网使用8B/10B、百兆以太网使用4B/5B），万兆传输标准如表1-4所示。

表1-4 万兆光纤传输标准

| 光纤协议 | 光纤标准 | 光纤尺寸/μm | 传输波长/nm | 传输距离 |
| --- | --- | --- | --- | --- |
| 10GBase-LX4 | 多模 | 50/125 | 1310 | 300 m |
|  | 单模 | 9/125 |  | 10 km |
| 10GBase-SR | 多模 | 62.5/125 | 850 | 33 m |
|  |  | 50/125 |  | 65m，新型2000MHz/km多模光纤上最长距离300 m |
| 10GBase-LR | 单模 | 9/125 | 1310 | 10 km |
| 10GBase-ER | 单模 | 9/125 | 1550 | 40 km |
| 10GBase-SW | 多模 | 62.5/125 | 850 | 33 m |
|  |  | 50/125 |  | 65 m，新型2000MHz/km多模光纤上最长距离300 m |
| 10GBase-LW | 单模 | 9/125 | 1310 | 10 km |
| 10GBase-EW | 单模 | 9/125 | 1550 | 40 km |

5）园区网设备接口的主要类型

（1）电接口类型

- RJ-45接口：常见设备的固化和模块化电口、Mini-GBIC-GT模块、GBIC-GT模块均为RJ-45接口。

- 同轴电缆：10GBase-CX4 模块，万兆铜缆接口。

（2）光接口类型
- SC 接口：GBIC-SX 模块、GBIC-LX 模块、10GBase-SR 模块、10GBase-LR 模块、10GBase-ER 模块均为 SC 接口。
- LC 接口：Mini-GBIC-SX 模块、Mini-GBIC-LX 模块、Mini-GBIC-LH 模块、Mini-GBIC-ZX50 模块、Mini-GBIC-ZX80 模块、Mini-GBIC-ZX100 模块均为 LC 接口。

SFP 接口均使用 Mini-GBIC 模块，光接口均为 LC 接口。

6）园区网常用模块

GBIC-GT 模块、Mini-GBIC-GT 模块、GBIC-SX 模块、GBIC-LX 模块、Mini-GBIC-SX 模块、Mini-GBIC-LX 模块。

很重要的一点：GBIC-GT 模块、Mini-GBIC-GT 模块的 1 000Mbit/s 接口，只能工作在 1 000Mbit/s 全双工模式下。

## 项目实施

1）项目设计

校园网建设中的第一个环节就是物理链路铺设，以主楼为中心，铺设光纤链路，覆盖周围的教学楼、实训楼、图书馆、综合楼、宿舍区等。以主楼为中心建设总配线间，以星形方式直接连接至各大楼。原则上，每一栋大楼设置一个主配线间，对于个别规模超大的楼宇根据需要，可以设计两个主配线间，如教学楼中设南主配线间和北主配线间。由主楼信息中心通过电信网和教育网接入外网，可参考图 1-2、图 1-3 所示。

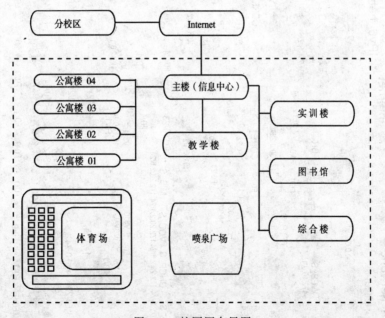

图 1-2　校园网布局图

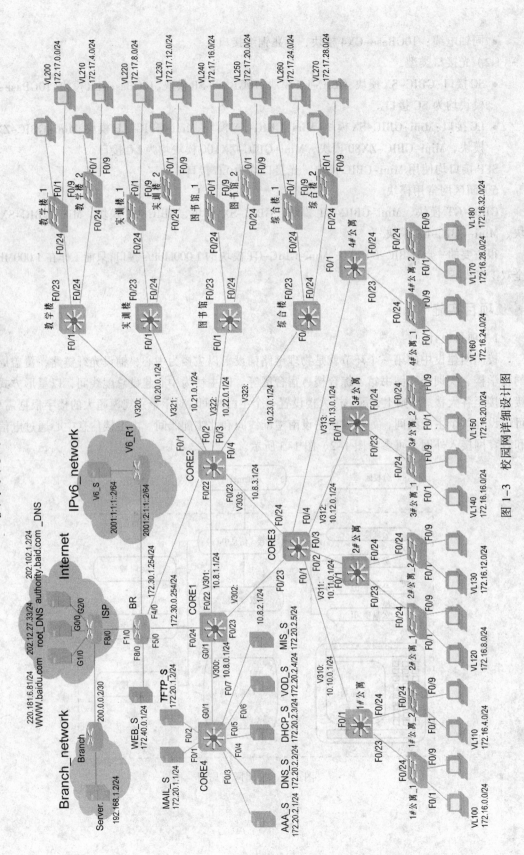

图1-3 校园网详细设计图

**注意**

本项目考虑到在 CISCO Packet Tracer5.3 上能完成,核心层和汇聚层均使用 3560 交换机,由于 3560 交换机接口数量有限,楼宇之间的光纤连接只能使用快速以太网接口进行,特此说明。

2)详细设计

① 详细设计请参考图 1-3。

② 打开 Packet Trace 5.3,在物理视图下修改园区名称为:CZIE(校园网),在逻辑视图下,分析表 1-5、表 1-6 和表 1-7,按照设备型号,添加、命名各设备(路由器、交换机、PC、Server)。

表 1-5 CZIE 园区宿舍区规划表

| 楼宇 | 设备间 | 设备名称 | 端口 | 类型 | 对端设备:端口 | 设备类型 |
|---|---|---|---|---|---|---|
| 1#公寓 | 1#公寓 | DS_1_GY | F0/1 | CF | CS_3:F0/1 | 3560 |
| | | | F0/23 | CF | AS_1_GY_1:F0/24 | |
| | | | F0/24 | CF | AS_1_GY_2:F0/24 | |
| | | AS_1_GY_1 | F0/24 | CF | | 2950-24 |
| | | | F0/1 | CF | PC0 | |
| | | | F0/9 | CF | PC1 | |
| | | PC0 | FE | CF | | PC |
| | | PC1 | FE | CF | | PC |
| | | AS_1_GY_2 | F0/24 | CF | | 2950-24 |
| | | | F0/1 | CF | PC2 | |
| | | | F0/9 | CF | PC3 | |
| | | PC2 | FE | CF | | PC |
| | | PC3 | FE | CF | | PC |
| 2#公寓 | 2#公寓 | DS_2_GY | F0/1 | CF | CS_3:F0/2 | 3560 |
| | | | F0/23 | CF | AS_2_GY_1:F0/24 | |
| | | | F0/24 | CF | AS_2_GY_2:F0/24 | |
| | | AS_2_GY_1 | F0/24 | CF | | 2950-24 |
| | | AS_1_GY_2 | F0/24 | CF | | 2950-24 |
| 3#公寓 | 3#公寓 | DS_3_GY | F0/1 | CF | CS_3:F0/3 | 3560 |
| | | | F0/23 | CF | AS_3_GY_1:F0/24 | |
| | | | F0/24 | CF | AS_3_GY_2:F0/24 | |
| | | AS_3_GY_1 | F0/24 | CF | | 2950-24 |
| | | AS_3_GY_2 | F0/24 | CF | | 2950-24 |
| 4#公寓 | 4#公寓 | DS_4_GY | F0/1 | CF | CS_3:F0/4 | 3560 |
| | | | F0/23 | CF | AS_4_GY_1:F0/24 | |
| | | | F0/24 | CF | AS_4_GY_2:F0/24 | |
| | | AS_4_GY_1 | F0/24 | CF | | 2950-24 |

续表

| 楼宇 | 设备间 | 设备名称 | 端口 | 类型 | 对端设备:端口 | 设备类型 |
|---|---|---|---|---|---|---|
| 4#公寓 | 4#公寓 | AS_4_GY_2 | F0/24 | CF | | 2950-24 |
| | | | F0/1 | CF | PC4 | |
| | | | F0/9 | CF | PC5 | |
| | | PC4 | FE | CF | | PC |
| | | PC5 | FE | CF | | PC |

表 1-6　CZIE 园区教学办公区规划表

| 楼宇 | 设备间 | 设备名称 | 端口 | 类型 | 对端设备:端口 | 设备类型 |
|---|---|---|---|---|---|---|
| 教学楼 | 教学楼 | DS_JX | F0/1 | CF | CS_2:F0/1 | 3560 |
| | | | F0/23 | CF | AS_JX_1:F0/24 | |
| | | | F0/24 | CF | AS_JX_2:F0/24 | |
| | | AS_JX_1 | F0/24 | CF | | 2950-24 |
| | | | F0/1 | CF | PC6 | |
| | | | F0/9 | CF | PC7 | |
| | | PC6 | FE | CF | | PC |
| | | PC7 | FE | CF | | PC |
| | | AS_JX_2 | F0/24 | CF | | 2950-24 |
| | | | F0/1 | CF | PC8 | |
| | | | F0/9 | CF | PC9 | |
| | | PC8 | FE | CF | | PC |
| | | PC9 | FE | CF | | PC |
| 实训楼 | 实训楼 | DS_SX | F0/1 | CF | CS_2:F0/2 | 3560 |
| | | | F0/23 | CF | AS_2_GY_1:F0/24 | |
| | | | F0/24 | CF | AS_2_GY_2:F0/24 | |
| | | AS_SX_1 | F0/24 | CF | | 2950-24 |
| | | AS_SX_2 | F0/24 | CF | | 2950-24 |
| 图书馆 | 图书馆 | DS_TSG | F0/1 | CF | CS_2:F0/3 | 3560 |
| | | | F0/23 | CF | AS_TSG_1:F0/24 | |
| | | | F0/24 | CF | AS_TSG_2:F0/24 | |
| | | AS_TSG_1 | F0/24 | CF | | 2950-24 |
| | | AS_TSG_2 | F0/24 | CF | | 2950-24 |
| 综合楼 | 综合楼 | DS_ZH | F0/1 | CF | CS_2:F0/4 | 3560 |
| | | | F0/23 | CF | AS_ZH_1:F0/24 | |
| | | | F0/24 | CF | AS_ZH_2:F0/24 | |
| | | AS_ZH_1 | F0/24 | CF | | 2950-24 |
| | | AS_ZH_2 | F0/24 | CF | | 2950-24 |
| | | | F0/1 | CF | PC10 | |
| | | | F0/9 | CF | PC11 | |
| | | PC10 | FE | CF | | PC |
| | | PC11 | FE | CF | | PC |

表 1-7  CZIE 园区信息中心规划表

| 楼宇 | 设备间 | 设备名称 | 端口 | 类型 | 对端设备:端口 | 设备类型 |
|---|---|---|---|---|---|---|
| 主楼（信息中心） | 主楼（信息中心） | CORE1 | F0/22 | CF | CORE2:F0/22 | 3560 |
| | | | F0/23 | CF | CORE3:F0/23 | |
| | | | F0/24 | CF | BR:F5/0 | |
| | | | G0/1 | CF | CORE4：G0/1 | |
| | | CORE2 | F0/1 | CF | DS_JX:F0/1 | 3560 |
| | | | F0/2 | CF | DS_SX:F0/1 | |
| | | | F0/3 | CF | DS_TSG:F0/1 | |
| | | | F0/4 | CF | DS_ZH:F0/1 | |
| | | | F0/22 | CF | CORE1：F0/22 | |
| | | | F0/23 | CF | CORE3：F0/24 | |
| | | | F0/24 | CF | BR：F4/0 | |
| | | CORE3 | F0/1 | CF | DS_1_GY:F0/1 | 3560 |
| | | | F0/2 | CF | DS_2_GY:F0/1 | |
| | | | F0/3 | CF | DS_3_GY:F0/1 | |
| | | | F0/4 | CF | DS_4_GY:F0/1 | |
| | | | F0/23 | CF | CORE1：F0/23 | |
| | | | F0/24 | CF | CORE2：F0/23 | |
| | | CORE4 | F0/1 | CF | EMAIL | 3560 |
| | | | F0/2 | CF | TFTP | |
| | | | F0/3 | CF | AAA | |
| | | | F0/4 | CF | DNS | |
| | | | F0/5 | CF | DHCP | |
| | | | F0/6 | CF | VOD | |
| | | | F0/7 | CF | MIS | |
| | | | G0/1 | CF | CORE1：G0/1 | |
| | | EMAIL | FE | CF | CORE4:F0/1 | Server |
| | | TFTP | FE | CF | CORE4:F0/2 | |
| | | AAA | FE | CF | CORE4:F0/3 | |
| | | DNS | FE | CF | CORE4:F0/4 | |
| | | DHCP | FE | CF | CORE4:F0/5 | |
| | | VOD | FE | CF | CORE4:F0/6 | |
| | | MIS | FE | CF | CORE4:F0/7 | |
| | | BR | F1/0 | CF | ISP:F8/0 | Router |
| | | | F4/0 | CF | CORE1:F0/24 | |
| | | | F5/0 | CF | CORE2:F0/24 | |
| | | | F8/0 | CF | WEB | |
| | | WEB | FE | CF | BR:F8/0 | Server |

③ 在物理视图下创建楼宇，在楼宇中创建管理设备间，将相应设备搬至各个设备间。

④ 在逻辑视图下，使用合适的线缆连接各设备间设备，保证物理上连接成功。

⑤ 在物理视图下创建园区，名称为：Internet（因特网）和 Branch（分校网），分析表 1-8，并创建相应楼宇、设备间。

⑥ 在逻辑视图下，添加、命名相应设备，将设备搬至对应设备间，连接 CZIE 园区和 Internet、Branch 园区。

⑦ 保存文件为：校园网_拓扑.pkt。

表 1-8　Internet 和 Branch 园区规划表

| 楼宇 | 设备间 | 设备名称 | 端口 | 类型 | 对端设备：端口 | 设备类型 |
| --- | --- | --- | --- | --- | --- | --- |
| Internet | Internet | ISP | F8/0 | CF | BR:F1/0 | Router |
| | | | G0/0 | CF | root_DNS:G0/0 | |
| | | | G1/0 | CF | www.baidu.com:G0/0 | |
| | | | G2/0 | CF | authority.baidu.com_DNS:G0/0 | |
| | | root_DNS | GE | CF | ISP：G0/0 | Server |
| | | authority.baidu.com_DNS | GE | CF | ISP：G2/0 | |
| | | www.baidu.com | GE | CF | ISP：G1/0 | |
| Branch | Branch | Branch | FE | FE | ISP | Router |
| | | | FE | FE | Server | |
| | | Server | FE | FE | Branch | Server |

# 项目 2　配置校园网

## 项目描述

在校园网物理连接的基础上，第二个环节就是联通整个网络，保证常用服务器的使用，如 DNS 服务、Web 服务等。作为网络实施工程师，通常采用工程化方法进行配置，并通过测试与调试以保证运行效果。

本项目提供工程化配置脚本，网络之间使用静态路由方式进行通信，因为应用服务部分的具体配置将在后面章节的项目中详细讨论，本项目中的 DNS 服务、Web 服务、DHCP 服务都已经实现配置。

本项目考虑的重点是全网运行的效果体验，所以各设备的配置保证全网联通尤为重要。

## 知识准备

CISCO 系列交换机、路由器所使用的操作系统是 IOS 或 COS（catalyst operating system）。IOS 使用较为广泛，该操作系统和路由器所使用的操作系统都基于相同的内核和 Shell。COS 的优点在于命令体系比较易用。利用操作系统所提供的命令，可实现对交换机的配置与管理。

CISCO IOS 操作系统具有以下特点：
① 支持通过命令行（CLI）或 Web 界面对交换机进行配置。
② 支持通过交换机的控制端口或 Telnet 会话登录、连接、访问交换机。
③ 提供用户模式和特权模式两种命令执行级别，并提供全局配置、线路配置、接口配置和 VLAN 配置等多种级别的配置模式，以允许用户对交换机的资源进行配置。

1）设备配置方法

对交换机、路由器的配置与管理一般都通过计算机进行，使用配置线把交换机的 CONSOLE 端口和计算机连接起来，把一台计算机配置成为相连交换机的仿真终端设备，这样就可以通过计算机配置和管理交换机的参数。

一般来说，CISCO 交换机可以通过四种方式进行配置。

（1）使用 PC 通过 CONSOLE 口对交换机进行配置和管理

新交换机在进行第一次配置时必须通过 CONSOLE 口访问交换机。计算机的串口和交换机的 CONSOLE 口是通过反转线（roll over）进行连接的，反转线的一端接在交换机的 CONSOLE 口上，另一端接到一个 DB9-RJ45 转接头上，DB9 则接到计算机的串口上。计算机和交换机连接好后，就可以使用各种各样的终端软件配置交换机了。

（2）通过 Telnet 命令对交换机进行远程管理

如果管理员不在交换机跟前，可以通过 Telnet 命令远程管理交换机，当然这需要预先在交换机上配置 IP 地址和密码，并保证管理员的计算机和交换机之间是 IP 可达的（简单来讲就是能 ping 通）。CISCO 交换机通常支持多人同时通过 Telnet 远程管理交换机，每一个用户称为一个虚拟终端（VTY）。第一个用户为 VTY 0，第二个用户为 VTY 1，以此类推，一般可达 VTY 4。

（3）通过 Web 对交换机进行远程管理

通过 Web 对交换机进行远程管理是指管理员通过网络在 IE 中输入 "http://IP" 地址，从而对交换机进行基于窗口方式的管理，也需要在交换机上设置 IP 地址及密码，并打开 Web 功能。

（4）通过 Ethernet 上的 SNMP 网管工作站对交换机进行管理

通过网管工作站进行配置，需要在网络中有至少一台运行 CISCOworks 及 CISCOview 等的网管工作站，还需要另外购买网管软件。

在以上四种管理交换机的方式中，后面三种方式都要连接网络，都会占用网络带宽，又称带内管理。交换机第一次使用时，必须采用第一种方式对交换机进行配置，这种方式并不占用网络的带宽，通过控制线连接交换机和计算机，又称带外管理。

2）IOS 工作模式及其切换

根据配置管理的功能不同，CISCO 交换机的工作模式可分为三种：用户模式、特权模式、配置模式（全局配置模式、线路配置模式、接口配置模式、VLAN 配置模式、路由配置模式），各种模式的切换方式见图 1-4。

（1）用户模式

当 PC 和交换机建立连接，配置好仿真终端时，首先处于用户模式。

在用户模式下，可以使用少量用户模式命令，命令的功能也受到一定限制，用户模式命令的操作结果不会被保存。

用户模式状态：hostname>。

（2）特权模式

要想在 CISCO 交换机上使用更多的命令，必须进入特权模式。

通常由用户模式进入特权模式时，必须输入进入特权模式的命令：enable。在特权模式下，用户可以使用所有的特权命令。

特权模式状态：hostname#。

（3）配置模式

通过 configure terminal 命令，可以由特权模式进入配置模式。在配置模式下，可以使用更多的命令来修改交换机的系统参数。

使用配置模式的命令可以对当前的配置进行修改。如果用户保存了配置信息，这些命令将被保存下来，并在系统重新启动时再次执行。要进入各种配置模式，首先必须进入全局配置模式。从全局配置模式出发，可以进入接口配置模式等各种配置子模式。图1-4所示为 IOS 的不同工作模式以及各模式之间的关系。

① 全局配置模式：在全局配置模式下，使用 vlan vlan-id 命令进入该模式。要返回全局配置模式，输入 exit 命令或按下【Ctrl+Z】组合键。要返回特权模式，使用 end 命令。

使用该模式可以配置 VLAN 参数。

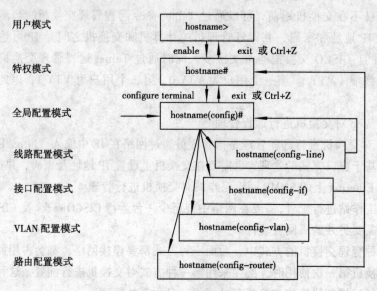

图 1-4　IOS 配置模式

② 线路配置模式：在全局配置模式下，执行 line VTY 或 line CONSOLE 命令，将进入线路配置模式。该模式主要用于对虚拟终端（VTY）和控制台（CONSOLE）进行配置，其主要是设置虚拟终端和控制台的用户级登录密码。

线路配置模式的命令行提示符为：hostname(config-line)#。

交换机有一个控制台，其编号为 0，通常利用该端口进行本地登录，以实现对交换机的配置和管理。

③ 接口配置模式：在全局配置模式下，执行 interface imterface-id 命令，将进入接口配置

模式。该模式主要用于对接口进行配置，主要是设置端口属性和 IP 地址。

接口配置模式的命令行提示符为：hostname(config-if)#。

④ VLAN 配置模式：在全局配置模式下，执行 vlan vlan-id 命令，将进入 VLAN 配置模式。该模式主要用于创建 VLAN 和命名 VLAN。

接口配置模式的命令行提示符为：hostname(config-vlan)#。

⑤ 路由配置模式：在全局配置模式下，执行 rouer rip|ospf|eigrp 命令，将进入路由配置模式。该模式主要用于配置各种动态路由协议。

接口配置模式的命令行提示符为：hostname(config-router)#。

## 项目实施

1）项目设计

按照校园网地址规划表（见表 1-9），配置各个设备，保证全网连通，测试 DHCP 等服务。

表 1-9　校园网地址规划表

| 楼宇 | 设备间 | 设备名称 | 端口 | IP 地址 | 子网掩码 | 网关\|VLAN |
|---|---|---|---|---|---|---|
| 1#公寓 | 1#公寓 | DS_1_GY | F0/1 | | | V310 |
| | | | F0/23 | | | V100 |
| | | | F0/24 | | | V110 |
| | | | V310 | 10.10.0.254 | 255.255.255.0 | |
| | | | V100 | 172.16.0.254 | 255.255.255.0 | |
| | | | V110 | 172.16.4.254 | 255.255.255.0 | |
| | | AS_1_GY_1 | F0/24 | | | |
| | | | F0/1 | | | |
| | | | F0/9 | | | |
| | | PC0 | FE | | | |
| | | PC1 | FE | | | |
| | | AS_1_GY_2 | F0/24 | | | |
| | | | F0/1 | | | |
| | | | F0/9 | | | |
| 1#公寓 | 1#公寓 | PC2 | FE | | | |
| | | PC3 | FE | | | |
| 2#公寓 | 2#公寓 | DS_2_GY | F0/1 | | | V311 |
| | | | F0/23 | | | V120 |
| | | | F0/24 | | | V130 |
| | | AS_2_GY_1 | F0/24 | | | |
| | | AS_1_GY_2 | F0/24 | | | |
| | | | V311 | 10.11.0.254 | 255.255.255.0 | |
| | | | V120 | 172.16.8.254 | 255.255.255.0 | |
| | | | V130 | 172.16.12.254 | 255.255.255.0 | |

续表

| 楼宇 | 设备间 | 设备名称 | 端口 | IP 地址 | 子网掩码 | 网关\|VLAN |
|---|---|---|---|---|---|---|
| 3#公寓 | 3#公寓 | DS_3_GY | F0/1 | | | V312 |
| | | | F0/23 | | | V140 |
| | | | F0/24 | | | V150 |
| | | AS_3_GY_1 | F0/24 | | | |
| | | AS_3_GY_2 | F0/24 | | | |
| | | | V312 | 10.12.0.254 | 255.255.255.0 | |
| | | | V140 | 172.16.16.254 | 255.255.255.0 | |
| | | | V150 | 172.16.20.254 | 255.255.255.0 | |
| 4#公寓 | 4#公寓 | DS_4_GY | F0/1 | | | V313 |
| | | | F0/23 | | | V160 |
| | | | F0/24 | | | V170/180 |
| | | AS_4_GY_1 | F0/24 | | | |
| | | AS_4_GY_2 | F0/24 | | | |
| | | | F0/1 | | | |
| | | | F0/9 | | | |
| | | PC4 | FE | | | |
| | | PC5 | FE | | | |
| | | | V313 | 10.13.0.254 | 255.255.255.0 | |
| | | | V160 | 172.16.24.254 | 255.255.255.0 | |
| | | | V170 | 172.16.28.254 | 255.255.255.0 | |
| | | | V180 | 172.16.32.254 | 255.255.255.0 | |
| 教学楼 | 教学楼 | DS_JX | F0/1 | | | V320 |
| | | | F0/23 | | | V200 |
| | | | F0/24 | | | V210 |
| | | AS_JX_1 | F0/24 | | | |
| | | | F0/1 | | | |
| | | | F0/9 | | | |
| | | PC6 | FE | | | |
| | | PC7 | FE | | | |
| | | AS_JX_2 | F0/24 | | | |
| | | | F0/1 | | | |
| | | | F0/9 | | | |
| 教学楼 | 教学楼 | PC8 | FE | | | |
| | | PC9 | FE | | | |
| | | | V320 | 10.20.0.254 | 255.255.255.0 | |
| | | | V200 | 172.17.0.254 | 255.255.255.0 | |
| | | | V210 | 172.17.4.254 | 255.255.255.0 | |

续表

| 楼宇 | 设备间 | 设备名称 | 端口 | IP 地址 | 子网掩码 | 网关|VLAN |
|---|---|---|---|---|---|---|
| 实训楼 | 实训楼 | DS_SX | F0/1 | | | V321 |
| | | | F0/23 | | | V220 |
| | | | F0/24 | | | V230 |
| | | AS_SX_1 | F0/24 | | | |
| | | AS_SX_2 | F0/24 | | | |
| | | | V321 | 10.21.0.254 | 255.255.255.0 | |
| | | | V220 | 172.17.8.254 | 255.255.255.0 | |
| | | | V230 | 172.17.12.254 | 255.255.255.0 | |
| 图书馆 | 图书馆 | DS_TSG | F0/1 | | | V323 |
| | | | F0/23 | | | V240 |
| | | | F0/24 | | | V250 |
| | | AS_TSG_1 | F0/24 | | | |
| | | AS_TSG_2 | F0/24 | | | |
| | | | V323 | 10.23.0.254 | 255.255.255.0 | |
| | | | V240 | 172.17.16.254 | 255.255.255.0 | |
| | | | V250 | 172.17.20.254 | 255.255.255.0 | |
| 综合楼 | 综合楼 | DS_ZH | F0/1 | | | V323 |
| | | | F0/23 | | | V260 |
| | | | F0/24 | | | V270 |
| | | AS_ZH_1 | F0/24 | | | |
| | | AS_ZH_2 | F0/24 | | | |
| | | | F0/1 | | | |
| | | | F0/9 | | | |
| | | PC10 | FE | | | |
| | | PC11 | FE | | | |
| | | | V323 | 10.23.0.254 | 255.255.255.0 | |
| | | | V260 | 172.17.24.254 | 255.255.255.0 | |
| | | | V270 | 172.17.28.254 | 255.255.255.0 | |
| (主楼)信息中心 | (主楼)信息中心 | CORE1 | F0/22 | | | V301 |
| | | | F0/23 | | | V302 |
| | | | F0/24 | 172.30.0.1 | | |
| | | | G0/1 | | | V300 |
| | | | V300 | 10.8.0.1 | | |
| | | | V301 | 10.8.1.1 | | |
| | | | V302 | 10.8.2.1 | | |
| | | CORE2 | F0/1 | | | V320 |

续表

| 楼宇 | 设备间 | 设备名称 | 端口 | IP 地址 | 子网掩码 | 网关\|VLAN |
|---|---|---|---|---|---|---|
| (主楼)信息中心 | (主楼)信息中心 | | F0/2 | | | V321 |
| | | | F0/3 | | | V322 |
| | | | F0/4 | | | V323 |
| | | | F0/22 | | | V301 |
| | | | F0/23 | | | V303 |
| | | | F0/24 | 172.30.1.1 | 255.255.255.0 | |
| | | | V301 | 10.8.1.254 | 255.255.255.0 | |
| | | | V303 | 10.8.3.1 | 255.255.255.0 | |
| | | | V320 | 10.20.0.1 | 255.255.255.0 | |
| | | | V321 | 10.21.0.1 | 255.255.255.0 | |
| | | | V322 | 10.22.0.1 | 255.255.255.0 | |
| | | | V323 | 10.23.0.1 | 255.255.255.0 | |
| | | CORE3 | F0/1 | | | V310 |
| | | | F0/2 | | | V311 |
| | | | F0/3 | | | V312 |
| | | | F0/4 | | | V313 |
| | | | F0/23 | | | V302 |
| | | | F0/24 | | | V303 |
| | | | V302 | 10.8.2.254 | 255.255.255.0 | |
| | | | V303 | 10.8.3.254 | 255.255.255.0 | |
| | | | V310 | 10.10.0.1 | 255.255.255.0 | |
| | | | V311 | 10.11.0.1 | 255.255.255.0 | |
| | | | V312 | 10.12.0.1 | 255.255.255.0 | |
| | | | V313 | 10.13.0.1 | 255.255.255.0 | |
| (主楼)信息中心 | (主楼)信息中心 | CORE4 | F0/1 | | | V500 |
| | | | F0/2 | | | V500 |
| | | | F0/3 | | | V501 |
| | | | F0/4 | | | V501 |
| | | | F0/5 | | | V501 |
| | | | F0/6 | | | V501 |
| | | | F0/7 | | | V501 |
| | | | G0/1 | | | V501 |
| | | | V500 | 172.20.1.254 | 255.255.255.0 | |
| | | | V501 | 172.20.2.254 | 255.255.255.0 | |
| | | EMAIL | FE | 172.20.1.1 | 255.255.255.0 | 172.20.1.254 |
| | | TFTP | FE | 172.20.1.2 | 255.255.255.0 | 172.20.1.254 |

续表

| 楼宇 | 设备间 | 设备名称 | 端口 | IP 地址 | 子网掩码 | 网关\|VLAN |
|---|---|---|---|---|---|---|
| （主楼）信息中心 | （主楼）信息中心 | AAA | FE | 172.20.2.1 | 255.255.255.0 | 172.20.2.254 |
| | | DNS | FE | 172.20.2.2 | 255.255.255.0 | 172.20.2.254 |
| | | DHCP | FE | 172.20.2.3 | 255.255.255.0 | 172.20.2.254 |
| | | VOD | FE | 172.20.2.4 | 255.255.255.0 | 172.20.2.254 |
| | | MIS | FE | 172.20.2.5 | 255.255.255.0 | 172.20.2.254 |
| | | BR | F0/0 | | | |
| | | | F1/0 | | | |
| | | | F4/0 | | | |
| | | | F5/0 | | | |
| | | | F8/0 | | | |
| | | WEB | FE | 172.40.0.1 | 255.255.255.0 | 172.40.0.254 |
| Internet | Internet | ISP | F8/0 | | | |
| | | | G0/0 | 202.12.27.1 | | |
| | | | G1/0 | 220.181.6.1 | | |
| | | | G2/0 | 202.102.1.1 | | |
| | | root_DNS | GE | 202.12.27.33 | 255.255.255.0 | 202.12.27.1 |
| | | authority.baidu.com_DNS | GE | 202.102.1.2 | 255.255.255.0 | 202.102.1.1 |
| | | www.baidu.com | GE | 220.181.6.81 | 255.255.255.0 | 220.181.6.1 |

2）项目实施

① 在 Packet Tracer 中打开"校园网_拓扑_Service_init.pkt"文件（文件由教师提供）或者"校园网_拓扑.pkt"文件（参考第 4 章内容自行配置 DHCP、DNS、WEB 服务），按照表 1-5 配置校园网地址。

② 核心层设备配置：按照表 1-10 配置：CORE1…CORE4。

表 1-10 核心层设备配置脚本

| CORE1 | CORE2 | CORE3 | CORE4 |
|---|---|---|---|
| configure terminal | configure terminal | configure terminal | configure terminal |
| hostname core1 | hostname core2 | hostname core3 | hostname Core4 |
| vlan 300 | vlan 301 | vlan 302 | vlan 300 |
|   name Core14 |   name Core12 |   name Core13 |   name Core14 |
|   exit |   exit |   exit |   exit |
| vlan 301 | vlan 303 | vlan 303 | vlan 500 |
|   name Core12 |   name Core23 |   name Core23 |   name ServerDMZ |
|   exit |   exit |   exit |   exit |
| vlan 302 | vlan 320 | vlan 310 | vlan 501 |
|   name Core13 |   name Teaching |   name GY1 |   name ServerINN |
|   exit |   exit |   exit |   exit |

续表

| CORE1 | CORE2 | CORE3 | CORE4 |
|---|---|---|---|
| interface FastEthernet0/22<br>  switchport mode access<br>  switchport access vlan 301<br>  exit<br>interface FastEthernet0/23<br>  switchport mode access<br>  switchport access vlan 302<br>  exit<br>interface FastEthernet0/24<br>  no switchport<br>  no shutdown<br>  ip address 172.30.0.1 255.255.255.0<br>  exit<br>interface GigabitEthernet0/1<br>  switchport mode access<br>  switchport access vlan 300<br>  exit<br>interface vlan 300<br>  no shutdown<br>  ip address 10.8.0.1 255.255.255.0<br>  exit<br>interface vlan 301<br>  no shutdown<br>  ip address 10.8.1.1 255.255.255.0<br>  exit<br>interface vlan 302<br>  no shutdown<br>  ip address 10.8.2.1 255.255.255.0<br>  exit<br>ip route 172.20.1.0 255.255.255.0 10.8.0.254<br>ip route 10.8.3.0 255.255.255.0 10.8.0.254<br>ip route 172.16.0.0 255.255.255.0 10.8.2.254 | vlan 321<br>  name Pracising<br>  exit<br>vlan 322<br>  name lib<br>  exit<br>vlan 323<br>  name common<br>  exit<br>interface FastEthernet0/1<br>  switchport mode access<br>  switchport access vlan 320<br>  exit<br>interface FastEthernet0/2<br>  switchport mode access<br>  switchport access vlan 321<br>  exit<br>interface FastEthernet0/3<br>  switchport mode access<br>  switchport access vlan 322<br>  exit<br>interface FastEthernet0/4<br>  switchport mode access<br>  switchport access vlan 323<br>  exit<br>interface FastEthernet0/22<br>  switchport mode access<br>  switchport access vlan 301<br>  exit<br>interface FastEthernet0/23<br>  switchport mode access<br>  switchport access vlan 303<br>  exit<br>interface FastEthernet0/24<br>  no switchport<br>  no shutdown<br>  ip address 172.30.1.1 255.255.255.0<br>  exit<br>interface vlan 301<br>  no shutdown<br>  ip address 10.8.1.254 | vlan 311<br>  name GY2<br>  exit<br>vlan 312<br>  name GY3<br>  exit<br>vlan 313<br>  name GY4<br>  exit<br>interface FastEthernet0/1<br>  switchport mode access<br>  switchport access vlan 310<br>  exit<br>interface FastEthernet0/2<br>  switchport mode access<br>  switchport access vlan 311<br>  exit<br>interface FastEthernet0/3<br>  switchport mode access<br>  switchport access vlan 312<br>  exit<br>interface FastEthernet0/4<br>  switchport mode access<br>  switchport access vlan 313<br>  exit<br>interface FastEthernet0/23<br>  switchport mode access<br>  switchport access vlan 302<br>  exit<br>interface FastEthernet0/24<br>  switchport mode access<br>  switchport access vlan 303<br>  exit<br>interface vlan 302<br>  no shutdown<br>  ip address 10.8.2.254 255.255.255.0<br>  exit<br>interface vlan 303<br>  no shutdown<br>  ip address 10.8.3.254 255.255.255.0 | interface Range FastEthernet0/1 – FastEthernet0/2<br>  switchport mode access<br>  switchport access vlan 500<br>  exit<br>interface Range FastEthernet0/3 – FastEthernet0/20<br>  switchport mode access<br>  switchport access vlan 501<br>  exit<br>interface GigabitEthernet0/1<br>  switchport mode access<br>  switchport access vlan 300<br>  exit<br>interface vlan 300<br>  no shutdown<br>  ip address 10.8.0.254 255.255.255.0<br>  exit<br>interface vlan 500<br>  no shutdown<br>  ip address 172.20.1.254 255.255.255.0<br>  exit<br>interface vlan 501<br>  no shutdown<br>  ip address 172.20.2.254 255.255.255.0<br>  exit<br>ip route 0.0.0.0 0.0.0.0 10.8.0.1<br>end<br>write |

续表

| CORE1 | CORE2 | CORE3 | CORE4 |
|---|---|---|---|
| ip route 10.10.0.0 255.255.255.0 10.8.2.254 | 255.255.255.0 exit | exit interface vlan 310 | |
| ip route 10.11.0.0 255.255.255.0 10.8.2.254 | interface vlan 303 no shutdown | no shutdown ip address 10.10.0.1 | |
| ip route 10.12.0.0 255.255.255.0 10.8.2.254 | ip address 10.8.3.1 255.255.255.0 | 255.255.255.0 exit | |
| ip route 10.13.0.0 255.255.255.0 10.8.2.254 | exit interface vlan 320 | interface vlan 311 no shutdown | |
| ip route 172.16.0.0 255.255.255.0 10.8.2.254 | no shutdown ip address 10.20.0.1 | ip address 10.11.0.1 255.255.255.0 | |
| ip route 172.16.4.0 255.255.255.0 10.8.2.254 | 255.255.255.0 exit | exit interface vlan 312 | |
| ip route 172.16.8.0 255.255.255.0 10.8.2.254 | interface vlan 321 no shutdown | no shutdown ip address 10.12.0.1 | |
| ip route 172.16.12.0 255.255.255.0 10.8.2.254 | ip address 10.21.0.1 255.255.255.0 | 255.255.255.0 exit | |
| ip route 172.16.16.0 255.255.255.0 10.8.2.254 | exit interface vlan 322 | interface vlan 313 no shutdown | |
| ip route 172.16.20.0 255.255.255.0 10.8.2.254 | no shutdown ip address 10.22.0.1 | ip address 10.13.0.1 255.255.255.0 | |
| ip route 172.16.24.0 255.255.255.0 10.8.2.254 | 255.255.255.0 exit | exit ip route 172.20.0.0 | |
| ip route 172.16.28.0 255.255.255.0 10.8.2.254 | interface vlan 323 no shutdown | 255.255.0.0 10.8.2.1 ip route 10.8.1.0 | |
| ip route 172.16.32.0 255.255.255.0 10.8.2.254 | ip address 10.23.0.1 255.255.255.0 | 255.255.255.0 10.8.2.1 ip route 10.20.0.0 | |
| ip route 10.20.0.0 255.255.255.0 10.8.1.254 | exit | 255.255.255.0 10.8.3.1 | |
| ip route 10.21.0.0 255.255.255.0 10.8.1.254 | ip route 172.20.1.0 255.255.255.0 10.8.1.1 | ip route 10.21.0.0 255.255.255.0 10.8.3.1 | |
| ip route 10.22.0.0 255.255.255.0 10.8.1.254 | ip route 172.20.2.0 255.255.255.0 10.8.1.1 | ip route 10.22.0.0 255.255.255.0 10.8.3.1 | |
| ip route 10.23.0.0 255.255.255.0 10.8.1.254 | ip route 10.8.2.0 255.255.255.0 10.8.1.1 | ip route 10.23.0.0 255.255.255.0 10.8.3.1 | |
| ip route 172.17.0.0 255.255.255.0 10.8.1.254 | ip route 10.10.0.0 255.255.255.0 10.8.3.254 | ip route 172.17.0.0 255.255.0.0 10.8.3.1 | |
| ip route 172.17.4.0 255.255.255.0 10.8.1.254 | ip route 10.11.0.0 255.255.255.0 10.8.3.254 | ip route 172.16.0.0 255.255.255.0 10.10.0.254 | |
| ip route 172.17.8.0 255.255.255.0 10.8.1.254 | ip route 10.12.0.0 255.255.255.0 10.8.3.254 | ip route 172.16.4.0 255.255.255.0 10.10.0.254 | |
| ip route 172.17.12.0 255.255.255.0 10.8.1.254 | ip route 10.13.0.0 255.255.255.0 10.8.3.254 | ip route 172.16.8.0 255.255.255.0 10.11.0.254 | |
| ip route 172.17.16.0 | ip route 172.16.0.0 | ip route 172.16.12.0 | |

续表

| CORE1 | CORE2 | CORE3 | CORE4 |
|---|---|---|---|
| 255.255.255.0 10.8.1.254 | 255.255.255.0 10.8.3.254 | 255.255.255.0 10.11.0.254 | |
| ip route 172.17.20.0 255.255.255.0 10.8.1.254 | ip route 172.16.4.0 255.255.255.0 10.8.3.254 | ip route 172.16.16.0 255.255.255.0 10.12.0.254 | |
| ip route 172.17.24.0 255.255.255.0 10.8.1.254 | ip route 172.16.8.0 255.255.255.0 10.8.3.254 | ip route 172.16.20.0 255.255.255.0 10.12.0.254 | |
| ip route 172.17.28.0 255.255.255.0 10.8.1.254 | ip route 172.16.12.0 255.255.255.0 10.8.3.254 | ip route 172.16.24.0 255.255.255.0 10.13.0.254 | |
| ip route 0.0.0.0 0.0.0.0 172.30.0.254 | ip route 172.16.16.0 255.255.255.0 10.8.3.254 | ip route 172.16.28.0 255.255.255.0 10.13.0.254 | |
| ip route 172.20.0.0 255.255.0.0 10.8.0.254 | ip route 172.16.20.0 255.255.255.0 10.8.3.254 | ip route 172.16.32.0 255.255.255.0 10.13.0.254 | |
| end | ip route 172.16.24.0 255.255.255.0 10.8.3.254 | ip route 0.0.0.0 0.0.0.0 10.8.2.1 | |
| write | ip route 172.16.28.0 255.255.255.0 10.8.3.254 | end | |
| | ip route 172.16.32.0 255.255.255.0 10.8.3.254 | write | |
| | ip route 172.17.0.0 255.255.255.0 10.20.0.254 | | |
| | ip route 172.17.4.0 255.255.255.0 10.20.0.254 | | |
| | ip route 172.17.8.0 255.255.255.0 10.21.0.254 | | |
| | ip route 172.17.12.0 255.255.255.0 10.21.0.254 | | |
| | ip route 172.17.16.0 255.255.255.0 10.22.0.254 | | |
| | ip route 172.17.20.0 255.255.255.0 10.22.0.254 | | |
| | ip route 172.17.24.0 255.255.255.0 10.23.0.254 | | |
| | ip route 172.17.28.0 255.255.255.0 10.23.0.254 | | |
| | ip route 172.20.0.0 255.255.0.0 10.8.1.1 | | |
| | ip route 0.0.0.0 0.0.0.0 172.30.1.254 | | |
| | end | | |
| | write | | |

③ 汇聚层配置：按照表1-11配置1#公寓…4#公寓。

表 1-11 宿舍楼汇聚层配置脚本

| 1#公寓 | 2#公寓 | 3#公寓 | 4#公寓 |
| --- | --- | --- | --- |
| configure terminal | configure terminal | configure terminal | configure terminal |
| hostname DS_1_GY | hostname DS_2_GY | hostname DS_3_GY | hostname DS_4_GY |
| vlan 310 | vlan 311 | vlan 312 | vlan 313 |
|   name GY1 |   name GY2 |   name GY3 |   name GY4 |
|   exit |   exit |   exit |   exit |
| vlan 100 | vlan 120 | vlan 140 | vlan 160 |
|   name Dorm1_1 |   name Dorm2_1 |   name Dorm3_1 |   name Dorm4_1 |
|   exit |   exit |   exit |   exit |
| vlan 110 | vlan 130 | vlan 150 | vlan 170 |
|   name Dorm1_2 |   name Dorm2_2 |   name Dorm3_2 |   name Dorm4_2 |
|   exit |   exit |   exit |   exit |
| interface FastEthernet0/1 | interface FastEthernet0/1 | interface FastEthernet0/1 | vlan 180 |
|   switchport mode access |   switchport mode access |   switchport mode access |   name Dorm4_3 |
|   switchport access vlan 310 |   switchport access vlan 311 |   switchport access vlan 312 |   exit |
|   exit |   exit |   exit | interface FastEthernet0/1 |
| interface FastEthernet0/23 | interface FastEthernet0/23 | interface FastEthernet0/23 |   switchport mode access |
|   no switchport |   switchport mode access |   switchport mode access |   switchport access vlan 313 |
|   ip address 172.16.0.254 255.255.255.0 |   switchport access vlan 120 |   switchport access vlan 140 |   exit |
|   ip helper-address 172.20.2.3 |   exit |   exit | interface FastEthernet0/23 |
|   exit | interface FastEthernet0/24 | interface FastEthernet0/24 |   switchport mode access |
| interface FastEthernet0/24 |   switchport mode access |   switchport mode access |   switchport access vlan 160 |
|   switchport mode access |   switchport access vlan 130 |   switchport access vlan 150 |   exit |
|   switchport access vlan 101 |   exit |   exit | interface FastEthernet0/24 |
|   exit | interface vlan 311 | interface vlan 312 |   switchport trunk encap dot1q |
| interface vlan 310 |   no shutdown |   no shutdown |   switchport mode trunk |
|   no shutdown |   ip address 10.11.0.254 255.255.255.0 |   ip address 10.12.0.254 255.255.255.0 |   exit |
|   ip address 10.10.0.254 255.255.255.0 |   exit |   exit | interface vlan 313 |
|   exit | interface vlan 120 | interface vlan 140 |   no shutdown |
| interface vlan 110 |   no shutdown |   no shutdown |   ip address 10.13.0.254 255.255.255.0 |
|   no shutdown |   ip address 172.16.8.254 255.255.255.0 |   ip address 172.16.16.254 255.255.255.0 |   exit |
|   ip address 172.16.4.254 255.255.255.0 |   exit |   exit | interface vlan 160 |
|   ip helper-address 172.20.2.3 | interface vlan 130 | interface vlan 150 |   no shutdown |
|   exit |   no shutdown |   no shutdown |   ip address 172.16.24.254 255.255.255.0 |
| interface FastEthernet0/24 |   ip address 172.16.12.254 255.255.255.0 |   ip address 172.16.20.254 255.255.255.0 |   exit |
|   switchport mode access |   exit |   exit | interface vlan 170 |
|   switchport access vlan 110 | ip route 0.0.0.0 0.0.0.0 10.11.0.1 | ip route 0.0.0.0 0.0.0.0 10.12.0.1 |   no shutdown |
|   exit | end | |   ip address 172.16.28.254 255.255.255.0 |

续表

| 1#公寓 | 2#公寓 | 3#公寓 | 4#公寓 |
|---|---|---|---|
| ip route 0.0.0.0 0.0.0.0 10.10.0.1<br>end<br>write | write | end<br>write | exit<br>interface vlan 180<br>no shutdown<br>ip address 172.16.32.254 255.255.255.0<br>exit<br>ip route 0.0.0.0 0.0.0.0 10.13.0.1<br>end<br>write |

④ 汇聚层配置：按照表1-12配置教学楼、实训楼、图书馆、综合楼。

表1-12 教学楼等汇聚层配置脚本

| 教学楼 | 实训楼 | 图书馆 | 综合楼 |
|---|---|---|---|
| configure terminal<br>hostname DS_JX<br>vlan 320<br>　name Teaching<br>　exit<br>vlan 200<br>　name Teach_1<br>　exit<br>vlan 210<br>　name Teach_2<br>　exit<br>interface FastEthernet0/1<br>　switchport mode access<br>　switchport access vlan 320<br>　exit<br>interface FastEthernet0/23<br>　switchport mode access<br>　switchport access vlan 200<br>　exit<br>interface FastEthernet0/24<br>　switchport mode access<br>　switchport access vlan 210<br>　exit<br>interface vlan 320<br>　no shutdown<br>　ip address 10.20.0.254 255.255.255.0 | configure terminal<br>hostname DS_SX<br>vlan 321<br>　name Practising<br>　exit<br>vlan 220<br>　name Practice_1<br>　exit<br>vlan 230<br>　name Practice_2<br>　exit<br>interface FastEthernet0/1<br>　switchport mode access<br>　switchport access vlan 321<br>　exit<br>interface FastEthernet0/23<br>　switchport mode access<br>　switchport access vlan 220<br>　exit<br>interface FastEthernet0/24<br>　switchport mode access<br>　switchport access vlan 230<br>　exit<br>interface vlan 321<br>　no shutdown<br>　ip address 10.21.0.254 255.255.255.0 | configure terminal<br>hostname DS_TSG<br>vlan 322<br>　name Lib<br>　exit<br>vlan 240<br>　name Lib_1<br>　exit<br>vlan 250<br>　name Lib_2<br>　exit<br>interface FastEthernet0/1<br>　switchport mode access<br>　switchport access vlan 322<br>　exit<br>interface FastEthernet0/23<br>　switchport mode access<br>　switchport access vlan 240<br>　exit<br>interface FastEthernet0/24<br>　switchport mode access<br>　switchport access vlan 250<br>　exit<br>interface vlan 322<br>　no shutdown<br>　ip address 10.22.0.254 255.255.255.0 | configure terminal<br>hostname DS_ZH<br>vlan 323<br>　name common<br>　exit<br>vlan 260<br>　name com_1<br>　exit<br>vlan 270<br>　name com_2<br>　exit<br>interface FastEthernet0/1<br>　switchport mode access<br>　switchport access vlan 323<br>　exit<br>interface FastEthernet0/23<br>　switchport mode access<br>　switchport access vlan 260<br>　exit<br>interface FastEthernet0/24<br>　switchport mode access<br>　switchport access vlan 270<br>　exit<br>interface vlan 323<br>　no shutdown<br>　ip address 10.23.0.254 255.255.255.0 |

续表

| 教学楼 | 实训楼 | 图书馆 | 综合楼 |
|---|---|---|---|
| exit<br>interface vlan 200<br>　no shutdown<br>　ip address 172.17.0.254 255.255.255.0<br>exit<br>interface vlan 210<br>　no shutdown<br>　ip address 172.17.4.254 255.255.255.0<br>exit<br>ip route 0.0.0.0 0.0.0.0 10.20.0.1<br>end<br>write | exit<br>interface vlan 220<br>　no shutdown<br>　ip address 172.17.8.254 255.255.255.0<br>exit<br>interface vlan 230<br>　no shutdown<br>　ip address 172.17.12.254 255.255.255.0<br>exit<br>ip route 0.0.0.0 0.0.0.0 10.21.0.1<br>end<br>write | exit<br>interface vlan 240<br>　no shutdown<br>　ip address 172.17.16.254 255.255.255.0<br>exit<br>interface vlan 250<br>　no shutdown<br>　ip address 172.17.20.254 255.255.255.0<br>exit<br>ip route 0.0.0.0 0.0.0.0 10.22.0.1<br>end<br>write | exit<br>interface vlan 260<br>　no shutdown<br>　ip address 172.17.24.254 255.255.255.0<br>exit<br>interface vlan 270<br>　no shutdown<br>　ip address 172.17.28.254 255.255.255.0<br>exit<br>ip route 0.0.0.0 0.0.0.0 10.23.0.1<br>end<br>write |

⑤ 接入层配置：按照表 1-13 配置宿舍楼。

表 1-13　宿舍楼接入配置

| 1#公寓_1 | 1#公寓_2 | 2#公寓_1 | 2#公寓_2 | 3#公寓_1 | 3#公寓_2 | 4#公寓_1 | 4#公寓_2 |
|---|---|---|---|---|---|---|---|
| configure terminal<br>hostname AS_1_GY_1<br>end<br>write | configure terminal<br>hostname AS_1_GY_2<br>vlan 110<br>name Dorm1_2<br>exit<br>interface range FastEthernet0/1-9<br>switchport mode access<br>switchport access vlan 130<br>exit<br>interface FastEthernet0/24<br>switchport mode trunk<br>exit | configure terminal<br>hostname AS_2_GY_1<br>vlan 120<br>name Dorm2_1<br>exit<br>interface range FastEthernet0/1-9<br>switchport mode access<br>switchport access vlan 130<br>exit<br>interface FastEthernet0/24<br>switchport mode trunk<br>exit | configure terminal<br>hostname AS_2_GY_2<br>vlan 130<br>name Dorm2_2<br>exit<br>interface range FastEthernet0/1-9<br>switchport mode access<br>switchport access vlan 130<br>exit<br>interface FastEthernet0/24<br>switchport mode trunk<br>exit | configure terminal<br>hostname AS_3_GY_1<br>vlan 140<br>name Dorm3_1<br>exit<br>interface range FastEthernet0/1-9<br>switchport mode access<br>switchport access vlan 140<br>exit<br>interface FastEthernet0/24<br>switchport mode trunk<br>exit | configure terminal<br>hostname AS_3_GY_2<br>vlan 150<br>name Dorm3_2<br>exit<br>interface range FastEthernet0/1-9<br>switchport mode access<br>switchport access vlan 150<br>exit<br>interface FastEthernet0/24<br>switchport mode trunk<br>exit | configure terminal<br>hostname AS_4_GY_1<br>vlan 160<br>name Dorm4_1<br>exit<br>interface range FastEthernet0/1-9<br>swichort mode access<br>switchport access vlan 160<br>exit<br>interface FastEthernet0/24<br>switchport mode trunk<br>exit | configure terminal<br>hostname AS_4_GY_2<br>vlan 170<br>name Dorm4_2<br>exit<br>vlan 180<br>name Dorm4_3<br>exit<br>interface range FastEthernet0/1-8<br>switchport mode access<br>switchport access vlan 170<br>exit<br>interface range |

续表

| 1#公寓_1 | 1#公寓_2 | 2#公寓_1 | 2#公寓_2 | 3#公寓_1 | 3#公寓_2 | 4#公寓_1 | 4#公寓_2 |
|---|---|---|---|---|---|---|---|
| end<br>write | end<br>write | end<br>write | end<br>write | end<br>write | end<br>write | end<br>write | FastEthernet0/9 - 16<br>switchport mode access<br>switchport access vlan 180<br>exit<br>interface FastEthernet0/24<br>switchport mode trunk<br>end<br>write |

⑥ 接入层配置：按照表1-14配置教学楼、实训楼、图书馆、综合楼。

表1-14 教学楼、实训楼、图书馆、综合楼接入配置

| | | | | | | | |
|---|---|---|---|---|---|---|---|
| configure terminal<br>hostname AS_JX_1<br>vlan 200<br> name Teach_1<br>exit<br>interface range FastEthernet0/1 - 9<br>switchport mode access<br>switchport access vlan 200<br>exit<br>interface FastEthernet0/24<br>switchport mode trunk<br>exit<br>end<br>write | configure terminal<br>hostname AS_JX2<br>vlan 210<br> name Teach_2<br>exit<br>interface range FastEthernet0/1 - 9<br>switchport mode access<br>switchport access vlan 210<br>exit<br>interface FastEthernet0/24<br>switchport mode trunk<br>exit<br>end<br>write | configure terminal<br>hostname AS_SX_1<br>vlan 220<br> name Practice_1<br>exit<br>interface range FastEthernet0/1 - 9<br>switchport mode access<br>switchport access vlan 220<br>exit<br>interface FastEthernet0/24<br>switchport mode trunk<br>exit<br>end<br>write | configure terminal<br>hostname AS_SX_2<br>vlan 230<br> name Practice_2<br>exit<br>interface range FastEthernet0/1 - 9<br>switchport mode access<br>switchport access vlan 230<br>exit<br>interface FastEthernet0/24<br>switchport mode trunk<br>exit<br>end<br>write | configure terminal<br>hostname AS_TSG_1<br>vlan 240<br> nameLib_1<br>exit<br>interface range FastEthernet0/1 - 9<br>switchport mode access<br>switchport access vlan 240<br>exit<br>interface FastEthernet0/24<br>switchport mode trunk<br>exit<br>end<br>write | configure terminal<br>hostname AS_TSG_2<br>vlan 250<br> nameLib_2<br>exit<br>interface range FastEthernet0/1 - 9<br>switchport mode access<br>switchport access vlan 250<br>exit<br>interface FastEthernet0/24<br>switchport mode trunk<br>exit<br>end<br>write | configure terminal<br>hostname AS_ZH_1<br>vlan 260<br> namecom_1<br>exit<br>interface range FastEthernet0/1 - 9<br>switchport mode access<br>switchport access vlan 260<br>exit<br>interface FastEthernet0/24<br>switchport mode trunk<br>exit<br>end<br>write | configure terminal<br>hostname AS_ZH_2<br>vlan 270<br> namecom_2<br>exit<br>interface range FastEthernet0/1 - 9<br>switchport mode access<br>switchport access vlan 270<br>exit<br>interface FastEthernet0/24<br>switchport mode trunk<br>exit<br>end<br>write |

⑦ 在 PC0 上使用 DHCP 方式获取地址，测试是否成功。
⑧ 在 PC0 上访问 www.czie.net ，测试是否成功。
⑨ 在 Packet Tracer 中另存文件为："校园网_拓扑_Service_S_Route.pkt"。

### 工程化操作

① 安装网络设备，保证物理接通。

② 创建核心层配置脚本。新建文本文件，将各个设备配置命令复制、粘贴到其中，保存文件名为：设备名_CFG.txt。

创建 CORE1 的脚本文件为 CORE1_CFG.txt，内容如下：

```
##################    core1-cfg    ##################
configure terminal
hostname core1
vlan 300
 name Core14
 exit
vlan 301
 name Core12
 exit
vlan 302
 name Core13
 exit
interface FastEthernet0/22
 switchport mode access
 switchport access vlan 301
 exit
interface FastEthernet0/23
 switchport mode access
 switchport access vlan 302
 exit
interface FastEthernet0/24
 no switchport
 no shutdown
 ip address 172.30.0.1 255.255.255.0
 exit
interface GigabitEthernet0/1
 switchport mode access
 switchport access vlan 300
 exit
interface vlan 300
 no shutdown
 ip address 10.8.0.1 255.255.255.0
 exit
interface vlan 301
 no shutdown
 ip address 10.8.1.1 255.255.255.0
 exit
interface vlan 302
 no shutdown
 ip address 10.8.2.1 255.255.255.0
 exit
```

```
 ip route 172.20.1.0 255.255.255.0 10.8.0.254
 ip route 10.8.3.0 255.255.255.0 10.8.0.254
 ip route 172.16.0.0 255.255.255.0 10.8.2.254
 ip route 10.10.0.0 255.255.255.0 10.8.2.254
 ip route 10.11.0.0 255.255.255.0 10.8.2.254
 ip route 10.12.0.0 255.255.255.0 10.8.2.254
 ip route 10.13.0.0 255.255.255.0 10.8.2.254
 ip route 172.16.0.0 255.255.255.0 10.8.2.254
 ip route 172.16.4.0 255.255.255.0 10.8.2.254
 ip route 172.16.8.0 255.255.255.0 10.8.2.254
 ip route 172.16.12.0 255.255.255.0 10.8.2.254
 ip route 172.16.16.0 255.255.255.0 10.8.2.254
 ip route 172.16.20.0 255.255.255.0 10.8.2.254
 ip route 172.16.24.0 255.255.255.0 10.8.2.254
 ip route 172.16.28.0 255.255.255.0 10.8.2.254
 ip route 172.16.32.0 255.255.255.0 10.8.2.254
 ip route 10.20.0.0 255.255.255.0 10.8.1.254
 ip route 10.21.0.0 255.255.255.0 10.8.1.254
 ip route 10.22.0.0 255.255.255.0 10.8.1.254
 ip route 10.23.0.0 255.255.255.0 10.8.1.254
 ip route 172.17.0.0 255.255.255.0 10.8.1.254
 ip route 172.17.4.0 255.255.255.0 10.8.1.254
 ip route 172.17.8.0 255.255.255.0 10.8.1.254
 ip route 172.17.12.0 255.255.255.0 10.8.1.254
 ip route 172.17.16.0 255.255.255.0 10.8.1.254
 ip route 172.17.20.0 255.255.255.0 10.8.1.254
 ip route 172.17.24.0 255.255.255.0 10.8.1.254
 ip route 172.17.28.0 255.255.255.0 10.8.1.254
 ip route 0.0.0.0 0.0.0.0 172.30.0.254
 ip route 172.20.0.0 255.255.0.0 10.8.0.254
 end
write
```

创建 CORE2 的脚本文件为 CORE2_CFG.txt，内容如下：

```
##################    core2-cfg    ##################
configure terminal
hostname core2
vlan 301
 name Core12
 exit
vlan 303
 name Core23
 exit
vlan 320
 name Teaching
 exit
vlan 321
 name Pracising
 exit
vlan 322
 name lib
 exit
vlan 323
 name common
```

```
 exit
interface FastEthernet0/1
 switchport mode access
 switchport access vlan 320
 exit
interface FastEthernet0/2
 switchport mode access
 switchport access vlan 321
 exit
interface FastEthernet0/3
 switchport mode access
 switchport access vlan 322
 exit
interface FastEthernet0/4
 switchport mode access
 switchport access vlan 323
 exit
interface FastEthernet0/22
 switchport mode access
 switchport access vlan 301
 exit
interface FastEthernet0/23
 switchport mode access
 switchport access vlan 303
 exit
interface FastEthernet0/24
 no switchport
 no shutdown
 ip address 172.30.1.1 255.255.255.0
 exit
interface vlan 301
 no shutdown
 ip address 10.8.1.254 255.255.255.0
 exit
interface vlan 303
 no shutdown
 ip address 10.8.3.1 255.255.255.0
 exit
interface vlan 320
 no shutdown
 ip address 10.20.0.1 255.255.255.0
 exit
interface vlan 321
 no shutdown
 ip address 10.21.0.1 255.255.255.0
 exit
interface vlan 322
 no shutdown
 ip address 10.22.0.1 255.255.255.0
 exit
interface vlan 323
 no shutdown
 ip address 10.23.0.1 255.255.255.0
 exit
ip route 172.20.1.0 255.255.255.0 10.8.1.1
```

```
ip route 172.20.2.0 255.255.255.0 10.8.1.1
ip route 10.8.2.0 255.255.255.0 10.8.1.1
ip route 10.10.0.0 255.255.255.0 10.8.3.254
ip route 10.11.0.0 255.255.255.0 10.8.3.254
ip route 10.12.0.0 255.255.255.0 10.8.3.254
ip route 10.13.0.0 255.255.255.0 10.8.3.254
ip route 172.16.0.0 255.255.255.0 10.8.3.254
ip route 172.16.4.0 255.255.255.0 10.8.3.254
ip route 172.16.8.0 255.255.255.0 10.8.3.254
ip route 172.16.12.0 255.255.255.0 10.8.3.254
ip route 172.16.16.0 255.255.255.0 10.8.3.254
ip route 172.16.20.0 255.255.255.0 10.8.3.254
ip route 172.16.24.0 255.255.255.0 10.8.3.254
ip route 172.16.28.0 255.255.255.0 10.8.3.254
ip route 172.16.32.0 255.255.255.0 10.8.3.254
ip route 172.17.0.0 255.255.255.0 10.20.0.254
ip route 172.17.4.0 255.255.255.0 10.20.0.254
ip route 172.17.8.0 255.255.255.0 10.21.0.254
ip route 172.17.12.0 255.255.255.0 10.21.0.254
ip route 172.17.16.0 255.255.255.0 10.22.0.254
ip route 172.17.20.0 255.255.255.0 10.22.0.254
ip route 172.17.24.0 255.255.255.0 10.23.0.254
ip route 172.17.28.0 255.255.255.0 10.23.0.254
ip route 172.20.0.0 255.255.0.0 10.8.1.1
ip route 0.0.0.0 0.0.0.0 172.30.1.254
 end
write
```

创建 CORE3 的脚本文件为 CORE3_CFG.txt, 内容如下:

```
################## core3-cfg ##################
configure terminal
hostname core3
vlan 302
 name Core13
 exit
vlan 303
 name Core23
 exit
vlan 310
 name GY1
 exit
vlan 311
 name GY2
 exit
vlan 312
 name GY3
 exit
vlan 313
 name GY4
 exit
exitinterface FastEthernet0/1
 switchport mode access
 switchport access vlan 310
 exit
```

```
interface FastEthernet0/2
 switchport mode access
 switchport access vlan 311
 exit
interface FastEthernet0/3
 switchport mode access
 switchport access vlan 312
 exit
interface FastEthernet0/4
 switchport mode access
 switchport access vlan 313
 exit
interface FastEthernet0/23
 switchport mode access
 switchport access vlan 302
 exit
interface FastEthernet0/24
 switchport mode access
 switchport access vlan 303
 exit
interface vlan 302
 no shutdown
 ip address 10.8.2.254 255.255.255.0
 exit
interface vlan 303
 no shutdown
 ip address 10.8.3.254 255.255.255.0
 exit
interface vlan 310
 no shutdown
 ip address 10.10.0.1 255.255.255.0
 exit
interface vlan 311
 no shutdown
 ip address 10.11.0.1 255.255.255.0
 exit
interface vlan 312
 no shutdown
 ip address 10.12.0.1 255.255.255.0
 exit
interface vlan 313
 no shutdown
 ip address 10.13.0.1 255.255.255.0
 exit
ip route 172.20.0.0 255.255.0.0 10.8.2.1
ip route 10.8.1.0 255.255.255.0 10.8.2.1
ip route 10.20.0.0 255.255.255.0 10.8.3.1
ip route 10.21.0.0 255.255.255.0 10.8.3.1
ip route 10.22.0.0 255.255.255.0 10.8.3.1
ip route 10.23.0.0 255.255.255.0 10.8.3.1
ip route 172.17.0.0 255.255.0.0 10.8.3.1
ip route 172.16.0.0 255.255.255.0 10.10.0.254
```

```
ip route 172.16.4.0 255.255.255.0 10.10.0.254
ip route 172.16.8.0 255.255.255.0 10.11.0.254
ip route 172.16.12.0 255.255.255.0 10.11.0.254
ip route 172.16.16.0 255.255.255.0 10.12.0.254
ip route 172.16.20.0 255.255.255.0 10.12.0.254
ip route 172.16.24.0 255.255.255.0 10.13.0.254
ip route 172.16.28.0 255.255.255.0 10.13.0.254
ip route 172.16.32.0 255.255.255.0 10.13.0.254
ip route 0.0.0.0 0.0.0.0 10.8.2.1
 end
write
```

创建 CORE4 的脚本文件为 CORE4_CFG.txt，内容如下：

```
##################       core4-cfg       ##################
configure terminal
hostname Core4
vlan 300
 name Core14
 exit
vlan 500
 name ServerDMZ
 exit
vlan 501
 name ServerINN
 exit
interface Range FastEthernet0/1 - FastEthernet0/2
 switchport mode access
 switchport access vlan 500
 exit
interface Range FastEthernet0/3 - FastEthernet0/20
 switchport mode access
 switchport access vlan 501
 exit
interface GigabitEthernet0/1
 switchport mode access
 switchport access vlan 300
 exit
interface vlan 300
 no shutdown
 ip address 10.8.0.254 255.255.255.0
 exit
interface vlan 500
 no shutdown
 ip address 172.20.1.254 255.255.255.0
 exit
interface vlan 501
 no shutdown
 ip address 172.20.2.254 255.255.255.0
 exit
ip route 0.0.0.0 0.0.0.0 10.8.0.1
 end
write
```

③ 创建汇聚层配置脚本。新建文本文件,将各个设备配置命令复制、粘贴到其中,保存文件名为:设备名_CFG.txt。

创建 DS_1_GY 的脚本文件为 DS_1_GY_CFG.txt,内容如下:

```
##################      ds-1-gy-cfg      ##################
configure terminal
hostname DS_1_GY
vlan 310
 name GY1
 exit
vlan 100
 name Dorm1_1
 exit
vlan 110
 name Dorm1_2
 exit
interface FastEthernet0/1
 switchport mode access
 switchport access vlan 310
 exit
interface FastEthernet0/23
 no switchport
 ip address 172.16.0.254 255.255.255.0
 exit
interface FastEthernet0/24
 switchport mode access
 switchport access vlan 101
 exit
interface vlan 310
 no shutdown
 ip address 10.10.0.254 255.255.255.0
 exit
interface vlan 110
 no shutdown
 ip address 172.16.4.254 255.255.255.0
 ip helper-address 172.20.2.3
 exit
interface FastEthernet0/24
 switchport mode access
 switchport access vlan 110
 exit
ip route 0.0.0.0 0.0.0.0 10.10.0.1
 end
write
```

创建 DS_2_GY 的脚本文件为 DS_2_GY_CFG.txt,内容如下:

```
##################      ds-2-gy-cfg      ##################
configure terminal
hostname DS_2_GY
vlan 311
```

```
 name GY2
 exit
vlan 120
 name Dorm2_1
 exit
vlan 130
 name Dorm2_2
 exit
interface FastEthernet0/1
 switchport mode access
 switchport access vlan 311
 exit
interface FastEthernet0/23
 switchport mode access
 switchport access vlan 120
 exit
interface FastEthernet0/24
 switchport mode access
 switchport access vlan 130
 exit
interface vlan 311
 no shutdown
 ip address 10.11.0.254 255.255.255.0
 exit
interface vlan 120
 no shutdown
 ip address 172.16.8.254 255.255.255.0
 exit
interface vlan 130
 no shutdown
 ip address 172.16.12.254 255.255.255.0
 exit
ip route 0.0.0.0 0.0.0.0 10.11.0.1
 end
write
```

创建 DS_3_GY 的脚本文件为 DS_3_GY_CFG.txt，内容如下：

```
################### ds-3-gy-cfg ###################
configure terminal
hostname DS_3_GY
vlan 312
 name GY3
 exit
vlan 140
 name Dorm3_1
 exit
vlan 150
 name Dorm3_2
 exit
interface FastEthernet0/1
```

```
 switchport mode access
 switchport access vlan 312
 exit
interface FastEthernet0/23
 switchport mode access
 switchport access vlan 140
 exit
interface FastEthernet0/24
 switchport mode access
 switchport access vlan 150
 exit
interface vlan 312
 no shutdown
 ip address 10.12.0.254 255.255.255.0
 exit
interface vlan 140
 no shutdown
 ip address 172.16.16.254 255.255.255.0
 exit
interface vlan 150
 no shutdown
 ip address 172.16.20.254 255.255.255.0
 exit
ip route 0.0.0.0 0.0.0.0 10.12.0.1
 end
write
```

创建 DS_4_GY 的脚本文件为 DS_4_GY_CFG.txt，内容如下：

```
##################    ds-4-gy-cfg    ##################
configure terminal
hostname DS_4_GY
vlan 313
 name GY4
 exit
vlan 160
 name Dorm4_1
 exit
vlan 170
 name Dorm4_2
 exit
interface FastEthernet0/1
 switchport mode access
 switchport access vlan 313
 exit
interface FastEthernet0/23
 switchport mode access
 switchport access vlan 160
 exit
interface FastEthernet0/24
 switchport mode access
 switchport access vlan 170
 exit
interface vlan 313
 no shutdown
```

```
 ip address 10.13.0.254 255.255.255.0
 exit
interface vlan 160
 no shutdown
 ip address 172.16.24.254 255.255.255.0
 exit
interface vlan 170
 no shutdown
 ip address 172.16.28.254 255.255.255.0
 exit
ip route 0.0.0.0 0.0.0.0 10.13.0.1
 end
write
```

创建 DS_JX 的脚本文件为 DS_JX_CFG.txt，内容如下：

```
##################    ds-jx-cfg    ##################
configure terminal
hostname DS_JX
vlan 320
 name Teaching
 exit
vlan 200
 name Teach_1
 exit
vlan 210
 name Teach_2
 exit
interface FastEthernet0/1
 switchport mode access
 switchport access vlan 320
 exit
interface FastEthernet0/23
 switchport mode access
 switchport access vlan 200
 exit
interface FastEthernet0/24
 switchport mode access
 switchport access vlan 210
 exit
interface vlan 320
 no shutdown
 ip address 10.20.0.254 255.255.255.0
 exit
interface vlan 200
 no shutdown
 ip address 172.17.0.254 255.255.255.0
 exit
interface vlan 210
 no shutdown
 ip address 172.17.4.254 255.255.255.0
 exit
ip route 0.0.0.0 0.0.0.0 10.20.0.1
 end
write
```

创建 DS_SX 的脚本文件为 DS_SX_CFG.txt，内容如下：

```
##################    ds-sx-cfg    ##################
configure terminal
hostname DS_SX
vlan 321
 name Practising
 exit
vlan 220
 name Practice_1
 exit
vlan 230
 name Practice_2
 exit
interface FastEthernet0/1
 switchport mode access
 switchport access vlan 321
 exit
interface FastEthernet0/23
 switchport mode access
 switchport access vlan 220
 exit
interface FastEthernet0/24
 switchport mode access
 switchport access vlan 230
 exit
interface vlan 321
 no shutdown
 ip address 10.21.0.254 255.255.255.0
 exit
interface vlan 220
 no shutdown
 ip address 172.17.8.254 255.255.255.0
 exit
interface vlan 230
 no shutdown
 ip address 172.17.12.254 255.255.255.0
 exit
ip route 0.0.0.0 0.0.0.0 10.21.0.1
 end
write
```

创建 DS_TSG 的脚本文件为 DS_TSG_CFG.txt，内容如下：

```
##################    ds-tsg-cfg    ##################
configure terminal
hostname DS_TSG
vlan 322
 name Lib
 exit
vlan 240
 name Lib_1
 exit
vlan 250
 name Lib_2
 exit
```

```
interface FastEthernet0/1
 switchport mode access
 switchport access vlan 322
 exit
interface FastEthernet0/23
 switchport mode access
 switchport access vlan 240
 exit
interface FastEthernet0/24
 switchport mode access
 switchport access vlan 250
 exit
interface vlan 322
 no shutdown
 ip address 10.22.0.254 255.255.255.0
 exit
interface vlan 240
 no shutdown
 ip address 172.17.16.254 255.255.255.0
 exit
interface vlan 250
 no shutdown
 ip address 172.17.20.254 255.255.255.0
 exit
ip route 0.0.0.0 0.0.0.0 10.22.0.1
 end
write
```

创建 DS_ZH 的脚本文件为 DS_ZH_CFG.txt，内容如下：

```
##################    ds-zh-cfg    ##################
configure terminal
hostname DS_ZH
vlan 323
 name common
 exit
vlan 260
 name com_1
 exit
vlan 270
 name com_2
 exit
interface FastEthernet0/1
 switchport mode access
 switchport access vlan 323
 exit
interface FastEthernet0/23
 switchport mode access
 switchport access vlan 260
 exit
interface FastEthernet0/24
 switchport mode access
 switchport access vlan 270
 exit
interface vlan 323
 no shutdown
```

```
 ip address 10.23.0.254 255.255.255.0
 exit
interface vlan 260
 no shutdown
 ip address 172.17.24.254 255.255.255.0
 exit
interface vlan 270
 no shutdown
 ip address 172.17.28.254 255.255.255.0
 exit
ip route 0.0.0.0 0.0.0.0 10.23.0.1
 end
write
```

④ 创建接入层配置脚本。新建文本文件，将各个设备配置命令复制、粘贴到其中并进行保存，文件名为：设备名_CFG.txt。

创建 AS_1_GY_1 的脚本文件为 AS_1_GY_1_CFG.txt，内容如下：

```
##################        as-1-gy-1-cfg      ##################
configure terminal
hostname  AS_1_GY_1
 end
write
```

创建 AS_1_GY_2 的脚本文件为 AS_1_GY_2_CFG.txt，内容如下：

```
##################        as-1-gy-2-cfg      ##################
configure terminal
hostname  AS_1_GY_2
vlan 110
 name Dorm1_2
 exit
interface range FastEthernet0/1 - 9
 switchport mode access
 switchport access vlan 130
 exit
interface FastEthernet0/24
 switchport mode trunk
  end
write
```

创建 AS_2_GY_1 的脚本文件为 AS_2_GY_1_CFG.txt，内容如下：

```
##################        as-2-gy-1-cfg      ##################
configure terminal
hostname  AS_2_GY_1
vlan 120
 name Dorm2_1
 exit
interface range FastEthernet0/1 - 9
 switchport mode access
 switchport access vlan 130
 exit
interface FastEthernet0/24
 switchport mode trunk
 end
write
```

创建 AS_2_GY_2 的脚本文件为 AS_2_GY_2_CFG.txt，内容如下：

```
##################      as-2-gy-2-cfg      ##################
configure terminal
hostname AS_2_GY_2
vlan 130
 name Dorm2_2
 exit
interface range FastEthernet0/1 - 9
 switchport mode access
 switchport access vlan 130
 exit
interface FastEthernet0/24
 switchport mode trunk
 end
write
```

创建 AS_3_GY_1 的脚本文件为 AS_3_GY_1_CFG.txt，内容如下：

```
##################      as-3-gy-1-cfg      ##################
configure terminal
hostname AS_3_GY_1
vlan 140
 name Dorm3_1
 exit
interface range FastEthernet0/1 - 9
 switchport mode access
 switchport access vlan 140
 exit
interface FastEthernet0/24
 switchport mode trunk
 end
write
```

创建 AS_3_GY_2 的脚本文件为 AS_3_GY_2_CFG.txt，内容如下：

```
##################      as-3-gy-2-cfg      ##################
configure terminal
hostname AS_3_GY_2
vlan 150
 name Dorm3_2
 exit
interface range FastEthernet0/1 - 9
 switchport mode access
 switchport access vlan 150
 exit
interface FastEthernet0/24
 switchport mode trunk
 end
write
```

创建 AS_4_GY_1 的脚本文件为 AS_4_GY_1_CFG.txt，内容如下：

```
##################      as-4-gy-1-cfg      ##################
configure terminal
hostname AS_4_GY_1
vlan 160
```

```
 name Dorm4_1
 exit
interface range FastEthernet0/1 - 9
 switchport mode access
 switchport access vlan 160
 exit
interface FastEthernet0/24
 switchport mode trunk
 end
write
```

创建 AS_4_GY_2 的脚本文件为 AS_4_GY_2_CFG.txt，内容如下：

```
##################     as-4-gy-2-cfg    ##################
configure terminal
hostname  AS_4_GY_2
vlan 170
 name 4_GY_2_1
 exit
vlan 180
 name 4_GY_2_2
 exit
interface range FastEthernet0/1 - 8
 switchport mode access
 switchport access vlan 170
 exit
interface range FastEthernet0/9 - 16
 switchport mode access
 switchport access vlan 180
 exit
interface FastEthernet0/24
 switchport mode trunk
 end
write
```

创建 AS_JX_1 的脚本文件为 AS_JX_1_CFG.txt，内容如下：

```
##################     as-jx-1-cfg     ##################
configure terminal
hostname  AS_JX_1
vlan 200
 name Teach_1
 exit
interface range FastEthernet0/1 - 9
 switchport mode access
 switchport access vlan 200
 exit
interface FastEthernet0/24
 switchport mode trunk
 end
write
```

创建 AS_JX_2 的脚本文件为 AS_JX_2_CFG.txt，内容如下：

```
##################     as-jx-2-cfg     ##################
configure terminal
hostname  AS_JX_2
```

```
vlan 210
 name Teach_2
 exit
interface range FastEthernet0/1 - 9
 switchport mode access
 switchport access vlan 210
 exit
interface FastEthernet0/24
 switchport mode trunk
 end
write
```

创建 AS_SX_1 的脚本文件为 AS_SX_1_CFG.txt，内容如下：

```
##################     as-sx-1-cfg     ##################
configure terminal
hostname  AS_SX_1
vlan 220
 name Practice_1
 exit
interface range FastEthernet0/1 - 9
 switchport mode access
 switchport access vlan 220
 exit
interface FastEthernet0/24
 switchport mode trunk
 end
write
```

创建 AS_SX_2 的脚本文件为 AS_SX_2_CFG.txt，内容如下：

```
##################     as-sx-2-cfg     ##################
configure terminal
hostname  AS_SX_2
vlan 230
 name Practice_2
 exit
interface range FastEthernet0/1 - 9
 switchport mode access
 switchport access vlan 230
 exit
interface FastEthernet0/24
 switchport mode trunk
 end
write
```

创建 AS_TSG_1 的脚本文件为 AS_TSG_1_CFG.txt，内容如下：

```
##################     as-tsg-1-cfg     ##################
configure terminal
hostname  AS_TSG_1
vlan 240
 name Lib_1
 exit
interface range FastEthernet0/1 - 9
 switchport mode access
 switchport access vlan 240
```

```
 exit
interface FastEthernet0/24
 switchport mode trunk
 end
write
```

创建 AS_TSG_2 的脚本文件为 AS_TSG_2_CFG.txt，内容如下：

```
##################     as-tsg-2-cfg     ##################
configure terminal
hostname AS_TSG_2
vlan 250
 name Lib_2
 exit
interface range FastEthernet0/1 - 9
 switchport mode access
 switchport access vlan 250
 exit
interface FastEthernet0/24
 switchport mode trunk
 end
write
```

创建 AS_ZH_1 的脚本文件为 AS_ZH_1_CFG.txt，内容如下：

```
##################     as-zh-1-cfg     ##################
configure terminal
hostname AS_ZH_1
vlan 260
 name com_1
 exit
interface range FastEthernet0/1 - 9
 switchport mode access
 switchport access vlan 260
 exit
interface FastEthernet0/24
 switchport mode trunk
 end
write
```

创建 AS_ZH_2 的脚本文件为 AS_ZH_2_CFG.txt，内容如下：

```
##################     as-zh-3-cfg     ##################
configure terminal
hostname AS_ZH_3
vlan 270
 name com_2
 exit
interface range FastEthernet0/1 - 9
 switchport mode access
 switchport access vlan 270
 exit
interface FastEthernet0/24
 switchport mode trunk
 end
write
```

⑤ 执行配置脚本：使用 CONSOLE 端口或者远程连接登录设备，打开超级终端或者 SecureCRT 程序，进入特权模式，复制设备中的命令脚本（设备名_CFG.txt），在超级终端或者 SecureCRT 程序中，右击并选择"粘贴"命令，然后执行脚本。

⑥ 验证配置：在各设备（如 CORE2）的特权模式下，使用 show running-config 命令检查当前运行配置，注意分析配置中各部分的内容，如表 1-15 所示。

表 1-15　当前运行配置

| 配　　置 | 备　　注 |
| --- | --- |
| core2#show running-config | 查看运行配置 |
| Building configuration... | 创建设备配置 |
| Current configuration : 2149 bytes | 当前配置大小：2149B |
| version 12.2 | 当前 IOS 版本：12.2 |
| no service timestamps log datetime msec | 关闭使用时间（精确到毫秒）标记日志信息服务 |
| no service timestamps debug datetime msec | 关闭使用时间（精确到毫秒）标记 debug 信息服务 |
| no service password-encryption | 关闭密码封装服务 |
| hostname core2 | 设备名：core2 |
| interface FastEthernet0/1<br>switchport access vlan 320<br>switchport mode access | F0/1 口<br>加入 VLAN 320<br>Access 模式 |
| interface FastEthernet0/24<br>no switchport<br>ip address 172.30.1.1 255.255.255.0<br>duplex auto<br>speed auto | F0/24 口<br>作为路由口使用<br>配置端口 IP 地址<br>Duplex 自适应<br>Speed 自适应 |
| interface Vlan1<br>no ip address<br>shutdown | VLAN 1<br>未配置 IP 地址<br>关闭 |
| ip route 172.20.1.0 255.255.255.0 10.8.1.1<br>ip route 172.20.2.0 255.255.255.0 10.8.1.1<br>ip route 10.8.2.0 255.255.255.0 10.8.1.1<br>ip route 10.10.0.0 255.255.255.0 10.8.3.254<br>ip route 10.11.0.0 255.255.255.0 10.8.3.254<br>ip route 10.12.0.0 255.255.255.0 10.8.3.254<br>ip route 10.13.0.0 255.255.255.0 10.8.3.254<br>ip route 172.16.0.0 255.255.255.0 10.8.3.254<br>ip route 172.16.4.0 255.255.255.0 10.8.3.254<br>ip route 172.16.8.0 255.255.255.0 10.8.3.254<br>ip route 172.16.12.0 255.255.255.0 10.8.3.254<br>ip route 172.16.16.0 255.255.255.0 10.8.3.254<br>ip route 172.16.20.0 255.255.255.0 10.8.3.254<br>ip route 172.16.24.0 255.255.255.0 10.8.3.254 | |

续表

| 配　　　　置 | 备　　　　注 |
|---|---|
| ip route 172.16.28.0 255.255.255.0 10.8.3.254 | 静态路由，可以检查目标网络与下一跳地址是否正确 |
| ip route 172.16.32.0 255.255.255.0 10.8.3.254 | |
| ip route 172.17.0.0 255.255.255.0 10.20.0.254 | |
| ip route 172.17.4.0 255.255.255.0 10.20.0.254 | |
| ip route 172.17.8.0 255.255.255.0 10.21.0.254 | |
| ip route 172.17.12.0 255.255.255.0 10.21.0.254 | |
| ip route 172.17.16.0 255.255.255.0 10.22.0.254 | |
| ip route 172.17.20.0 255.255.255.0 10.22.0.254 | |
| ip route 172.17.24.0 255.255.255.0 10.23.0.254 | |
| ip route 172.17.28.0 255.255.255.0 10.23.0.254 | |
| ip route 172.20.0.0 255.255.0.0 10.8.1.1 | |
| ip route 0.0.0.0 0.0.0.0 172.30.1.254 | 默认网络，用于访问路由表之外的网络 |
| ip classless | 启用无类路由，现行设备默认开启 |
| ip route 0.0.0.0 0.0.0.0 172.30.1.254 | 默认路由 |
| line con 0 | CONSOLE 口 |
| line vty 0 4 | VTY 线路（5 条） |
| 　login | 登录密码：xzchen |
| 　password xzchen | |
| end | 结束 |

⑦ 在 1#公寓楼最左端的 PC 的上使用 DHCP 方式获取地址。

⑧ 测试网络联通性：在设备上使用 ping 命令测试目标网络联通性。

⑨ 确认无误后，在设备上保存配置，并保存设备配置脚本文件。

# 本章训练内容

1. 完成第 8 章项目 17 中任务 1。
2. 完成第 8 章项目 18 中任务 1。

# 第 2 章　使用 VLAN 部署校园网

当前，交换技术迅速发展，也加快了虚拟局域网（VLAN：virtual local area network）技术的应用速度，特别是在校园网上的应用更广泛。通过将校园网络划分为虚拟子网，可以强化网络管理和网络安全，控制不必要的数据广播。数据广播在网络中起着非常重要的作用，随着校园网内计算机数量的增加，VOD 视频点播在课堂教学上的大量应用，广播包的数量也会急剧增加，当广播包的数量占到总量的 30%时，网络的传输速率将会明显下降。特别是当某网络设备出现故障后，会不停地向网络发送广播，从而导致网络风暴，使网络通信陷于瘫痪。当校园网络内计算机数超过 200 台后，建议人为分隔广播域。分隔广播域的方式有两种，一种是物理分隔，即将一个完整的网络物理地一分为二或一分为多，然后通过一个能够隔离广播的网络设备将彼此连接起来；另外一种是逻辑分隔，即将一个大的网络划分为若干个小的虚拟子络，也就是 VLAN，各 VLAN 间通过路由设备连接实现通信。

通过本章所有项目的实践，可以学会校园网中 VLAN 的规划、部署，通过配置接入 VLAN、汇聚 VLAN、管理 VLAN，提高网络的可利用性和可管理性。

本章需要完成的项目有：

项目 3——部署接入 VLAN；

项目 4——部署汇聚 VLAN；

项目 5——部署管理 VLAN。

## 项目 3　部署接入 VLAN

### 项目描述

随着校园信息化建设的发展，校园网的用户规模近万人，如果将所有用户都纳入同一网段，会带来很多问题：

- 广播流量太大，网络利用率降低。
- 网络 ARP 攻击导致安全性降低。
- 不便于对特定用户群（如部门）的管理。

为了有效地建设校园网，需要对校园网楼宇及部门用户需求进行分析，如何设计 VLAN？如何考虑校园网的 VLAN 升级？此类问题将在本项目中进行分析、解决。

通过本项目，读者可以掌握以下技能。

① 能够在接入层交换机配置创建、命名 VLAN。
② 能够将端口加入特定 VLAN。
③ 能够对端口和 VLAN 关系进行调整。
④ 能够对 VLAN 进行编址规划。

 知识准备

VLAN（virtual local area network）建立在局域网交换机的基础之上，是采用网络管理软件构建的可跨越不同网段、不同网络的端到端的逻辑网络。一个 VLAN 组成一个逻辑子网，即逻辑广播域，它可覆盖多个网络设备，允许处于不同地理位置的网络用户加入一个逻辑子网中，在功能和操作上与传统 LAN 基本相同，可提供一定范围内终端系统的互联。

1）VLAN 技术的特点

基于 VLAN 的校园网与普通校园网相比具有以下特点：

① 结点的增加、移动及其他变化都通过管理控制台快速、便捷处理，而无需到布线室进行调整。

② 通过限制整个网络的结点对结点通信和广播通信，虚拟子网可以节约带宽，从而降低成本。

③ 虚拟子网只能通过三层设备通信，因此需要时，基于路由器的标准安全措施可用于限制对每个虚拟子网的访问。

2）VLAN 的划分方式

（1）基于端口划分 VLAN

以网络设备的端口所属的具体 VLAN 的标准划分 VLAN。例如，在一个 8 端口的交换机中，端口 1、2、4、7 属于 VLAN1，而端口 3、5、6、8 属于 VLAN2，与这些端口相连的设备以及收发的数据帧相应地也分属于 VLAN1 和 VLAN2。这种划分方法的优点是：定义 VLAN 成员非常简单，指定所有端口即可；而其缺点是：不允许用户随便移动。如果属于某个 VLAN 的用户离开原来的端口，到达一个新网络设备的某端口，那么网络管理者必须重新定义该 VLAN。

（2）基于 MAC 地址划分 VLAN

以计算机工作站网卡的 MAC 地址划分 VLAN。这种划分方法的最大优点就是：由于一个 MAC 地址唯一对应一块网卡，因此当某个计算机工作站的物理位置改变时，即从一台交换机转移到另一台交换机时，无需重新配置 VLAN。

（3）基于网络层协议类型划分 VLAN

如果网络支持多网络层协议，那么根据数据帧头中的网络层协议类型字段也可划分 VLAN，这对网络管理者来说非常重要，同时这种方式无需附加帧标签识别 VLAN，可减少网络通信量。

3）基于端口 VLAN 的配置命令

IOS 中关于 VLAN 的常规配置命令如表 2-1 所示

表 2-1  VLAN 的配置命令

| 命 令 | 说 明 |
| --- | --- |
| vlan vlan-id | 创建 VLAN |

续表

| 命令 | 说明 |
|---|---|
| name vlan-name | 命名 VLAN |
| switchport mode access | 定义为 Access 模式 |
| switchport mode dynamic auto | 自动协商是否成为 trunk |
| switchport mode dynamic desirable | 如果对方端口是 trunk、desirable 或 auto 把端口设置为 trunk |
| switchport mode trunk | 设置端口为强制的 trunk，而不理会对方端口是否为 trunk |
| switchport trunk allowed vlan add vlan-id | 在 trunk 中添加 VLAN |
| switchport trunk allowed vlan remove vlan-id | 在 trunk 中移除 VLAN |
| switchport trunk allowed vlan except vlan-id | 在 trunk 中允许特定 VLAN 之外的其他所有 VLAN |
| switchport trunk allowed vlan all | 在 trunk 运行所有 VLAN |
| switchport trunk encapsulation isl | 使用 isl trunk 封装协议 |
| switchport trunk encapsulation dot1q | 使用 802.1q trunk 封装协议 |
| switchport access vlan vlan-id | 将端口划给特定 VLAN |
| switchport trunk native vlan vlan-id | 配置 native，默认为 VLAN 1 |
| show interfaces interface-id switchport | 显示有关 switchport 的配置 |
| show interfaces interface-id trunk | 显示有关 trunk 的配置 |
| show vlan | 显示 VLAN 和端口 |
| show vlan id vlan-id | 显示特定 VLAN |

## 项目实施

**1）项目设计**

接入 VLAN 主要是在接入层交换机上划分网络，有效减小广播域的范围，以确定所连接终端属于哪个部门。校园网楼宇与部门设计，请参考表 2-2。

表 2-2 楼宇与部门设计

| 楼宇 | 部门 | 设备 | VLAN ID |
|---|---|---|---|
| 1#公寓 | A 座 | AS_1_GY_1 | |
| | B 座 | AS_1_GY_2 | |
| 2#公寓 | A 座 | AS_2_GY_1 | |
| | B 座 | AS_2_GY_2 | |
| 3#公寓 | A 座 | AS_3_GY_1 | |
| | B 座 | AS_3_GY_2 | |
| 4#公寓 | A 座 | AS_4_GY_1 | |
| | B 座（1~3 层） | AS_4_GY_2 | |
| | B 座（4~6 层） | | |
| 教学楼 | A 楼 | AS_JX_1 | |
| | B 楼 | AS_JX_2 | |

续表

| 楼宇 | 部门 | 设备 | VLAN ID |
|---|---|---|---|
| 实训楼 | A 楼 | AS_SX_1 | |
| | B 楼 | AS_SX_2 | |
| 图书馆 | 教师部 | AS_TSG_1 | |
| | 读者部 | AS_TSG_2 | |
| 综合楼 | 财务部 | AS_ZH_1 | |
| | 行政部 | AS_ZH_2 | |

2）项目任务

按照表 2-2 中参数为各个部门创建 VLAN，命名 VLAN，将相应端口划入特定 VLAN，并进行验证。

> 注意
> 项目讲解以 4#公寓楼为例，请结合图 2-1，其他类似进行配置即可。

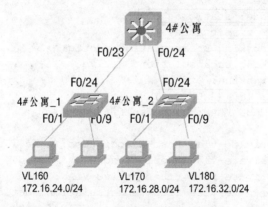

图 2-1　4#公寓楼接入图

① 在 AS_4_GY_2 交换机上，创建 VLAN 170、VLAN180，分别命名为 4_GY_2_1、4_GY_2_2。

```
AS_4_GY_2 (config)#vlan 170
AS_4_GY_2 (config-vlan)#name 4_GY_2_1
AS_4_GY_2 (config-vlan)#exit
AS_4_GY_2 (config)#vlan 180
AS_4_GY_2 (config-vlan)#name 4_GY_2_2
AS_4_GY_2 (config-vlan)#exit
```

② 在 AS_4_GY_2 交换机上，将 1～16 口设置为 Access 模式；将 1～8 口划入 VLAN170，9～16 口划入 VLAN180。

```
AS_4_GY_2 (config)#interface range fastethernet 0/1 - 8
AS_4_GY_2 (config-range-if)#switchport mode access
AS_4_GY_2 (config-range-if)#switchport access vlan 170
AS_4_GY_2 (config-range-if)#exit
AS_4_GY_2 (config)#interface range fastethernet 0/9 - 16
AS_4_GY_2 (config-range-if)#switchport mode access
```

```
AS_4_GY_2 (config-range-if)#switchport access vlan 180
AS_4_GY_2 (config-range-if)#exit
```

③ 通过 show vlan 命令查看 VLAN 端口。

```
AS_4_GY_2#show vlan
VLAN Name                         Status      Ports
-----------------------------------------------------------------------
1    default                      active      Fa0/17, Fa0/18, Fa0/19, Fa0/20
                                              Fa0/21, Fa0/22, Fa0/23
170  4_GY_2_1                     active      Fa0/1, Fa0/2, Fa0/3, Fa0/4
                                              Fa0/5, Fa0/6, Fa0/7, Fa0/8
180  4_GY_2_2                     active      Fa0/9, Fa0/10, Fa0/11, Fa0/12
                                              Fa0/13, Fa0/14, Fa0/15, Fa0/16
<省略部分>
```

④ 通过 show interface fastethernet fa0/* switchport 命令查看端口二层信息。

```
AS_4_GY_2#show interfaces fa0/1 switchport
Name: Fa0/1
Switchport: Enabled
Administrative Mode: static access
Operational Mode: static access
Administrative Trunking Encapsulation: dot1q
Operational Trunking Encapsulation: native
Negotiation of Trunking: Off
Access Mode VLAN: 170 (4_GY_2_1)
Trunking Native Mode VLAN: 1 (default)
Voice VLAN: none
Administrative private-vlan host-association: none
Administrative private-vlan mapping: none
Administrative private-vlan trunk native VLAN: none
Administrative private-vlan trunk encapsulation: dot1q
Administrative private-vlan trunk normal VLANs: none
Administrative private-vlan trunk private VLANs: none
Operational private-vlan: none
Trunking VLANs Enabled: ALL
Pruning VLANs Enabled: 2-1001
Capture Mode Disabled
Capture VLANs Allowed: ALL
Protected: false
Appliance trust: none
```

⑤ 可以看到 F0/1 端口的 Administrative Mode、Operational Mode 均为 static access，Access Mode VLAN 为 170。

⑥ 设计所有接入 VLAN，详见表 2-3。

表 2-3　VLAN 规划表

| 楼宇 | 部门 | 设　备 | VLAN ID | VLAN NAME | 端　口 |
|---|---|---|---|---|---|
| 1#公寓 | A 座 | AS_1_GY_1 | 100 | 1_GY_1 | F0/1-16 |
|  | B 座 | AS_1_GY_2 | 110 | 1_GY_2 | F0/1-16 |

续表

| 楼 宇 | 部 门 | 设 备 | VLAN ID | VLAN NAME | 端 口 |
|---|---|---|---|---|---|
| 2#公寓 | A座 | AS_2_GY_1 | 120 | 2_GY_1 | F0/1-16 |
| | B座 | AS_2_GY_2 | 130 | 2_GY_2 | F0/1-16 |
| 3#公寓 | A座 | AS_3_GY_1 | 140 | 3_GY_1 | F0/1-16 |
| | B座 | AS_3_GY_2 | 150 | 3_GY_2 | F0/1-16 |
| 4#公寓 | A座 | AS_4_GY_1 | 160 | 4_GY_1 | F0/1-16 |
| | B座（1~3层） | AS_4_GY_2 | 170 | 4_GY_2_1 | F0/1-8 |
| | B座（4~6层） | | 180 | 4_GY_2_2 | F0/9-16 |
| 教学楼 | A楼 | AS_JX_1 | 200 | JX_1 | F0/1-16 |
| | B楼 | AS_JX_2 | 210 | JX_2 | F0/1-16 |
| 实训楼 | A楼 | AS_SX_1 | 220 | SX_1 | F0/1-16 |
| | B楼 | AS_SX_2 | 230 | SX_2 | F0/1-16 |
| 图书馆 | 教师部 | AS_TSG_1 | 240 | Teach | F0/1-16 |
| | 读者部 | AS_TSG_2 | 250 | Stu | F0/1-16 |
| 综合楼 | 财务部 | AS_ZH_1 | 260 | Finan | F0/1-16 |
| | 行政部 | AS_ZH_2 | 270 | Admin | F0/1-16 |

 **注意**

考虑到校园网的升级，相邻的 VLAN ID 之间都相差 10，以便 VLAN 的扩展。

⑦ 请严格按照表 2-3 进行规划，在相应设备上配置 VLAN，将端口划入特定的 VLAN。

## 工程化操作

① 安装网络设备，保证物理联通。

② 创建配置脚本。新建文本文件，将各个设备配置命令复制、粘贴到其中，保存文件名为：设备名_ Acc VL_CFG.txt。例如，AS_4_GY_2 的脚本文件为 AS_4_GY_2_ Acc VL_CFG.txt，内容如下：

```
##################    as-4-gy-2- acc vl_cfg    ##################
configure terminal
hostname AS_4_GY_2
vlan 170
 name 4_GY_2_1
 exit
vlan 180
 name 4_GY_2_2
 exit
interface range FastEthernet0/1 - 8
 switchport mode access
 switchport access vlan 170
 exit
interface range FastEthernet0/9 - 16
 switchport mode access
```

```
 switchport access vlan 180
 exit
end
write
```

宿舍区、办公区其他接入交换机可以类似进行配置。例如，AS_1_GY_1 的脚本文件为 AS_1_GY_1_ Acc VL_CFG.txt，内容如下：

```
###################    as-1-gy-1- acc vl_cfg    ##################
configure terminal
hostname AS_1_GY_1
vlan 100
 name 1_GY_1
 exit
interface range FastEthernet0/1 - 16
 switchport mode access
 switchport access vlan 100
 exit
end
write
```

如果交换机中仅仅有一个 VLAN，可以不在本机创建，可以修改 AS_1_GY_1_ Acc VL_CFG.txt，内容如下：

```
###################    as-1-gy-1- acc vl_cfg    ##################
configure terminal
hostname AS_1_GY_1
end
write
```

但从交换机标准配置来看，建议配置 VLAN，将端口划入 VLAN。

③ 使用 CONSOLE 端口超级终端或者远程连接登录设备，进入特权模式，复制设备中的命令脚本，在终端上右击选择"粘贴"命令，执行脚本，检查是否有错误命令。

④ 使用 show vlan 命令验证 VLAN 和端口关系，如表 2-4 所示。

表 2-4  VLAN 和端口关系

| 配置 | | | 备注 |
|---|---|---|---|
| AS_1_GY_2#show vlan | | | |
| VLAN Name | Status | Ports | 交换机默认有一个 VLAN，编号为 1，不能删除，可以重命名，F0/1-16 口加入 VLAN 110（1_GY_2）中 |
| 1    default | active | Fa0/17, Fa0/18, Fa0/19, Fa0/20 Fa0/21, Fa0/22, Fa0/23 | |
| 110  1_GY_2 | active | Fa0/1, Fa0/2, Fa0/3, Fa0/4 Fa0/5, Fa0/6, Fa0/7, Fa0/8 Fa0/9, Fa0/10, Fa0/11, Fa0/12 Fa0/13, Fa0/14, Fa0/15, Fa0/16 | |
| 1002 fddi-default | act/unsup | | 默认的其他类型 VLAN |
| 1003 token-ring-default | act/unsup | | |
| 1004 fddinet-default | act/unsup | | |
| 1005 trnet-default | act/unsup | | |

续表

| 配置 | 备注 |
|---|---|
| VLAN Type SAID    MTU Parent RingNo BridgeNo Stp    BrdgMode Trans1 Trans2<br>---- ----- ---------- ----- ------ ------ --------- ---- -------- ------<br>1    enet  100001     1500  -      -      -         -    0        0<br>110  enet  100110     1500  -      -      -         -    0        0<br>&lt;省略部分&gt; | Type：VLAN 类型；<br>enet 为以太网；<br>MTU：最大传输单元，默认为 1500 |

⑤ 使用 show vlan id 110 命令查看 VLAN110 和端口关系。

⑥ 使用 show running-config 命令查看设备运行配置，如表 2-5 所示。

表 2-5　设备进行配置

| 配置 | 备注 |
|---|---|
| AS_1_GY_2#show running-config | 查看运行配置 |
| Building configuration...<br>Current configuration : 1637 bytes<br>!<br>version 12.1<br>no service timestamps log datetime msec<br>no service timestamps debug datetime msec<br>no service password-encryption<br>!<br>hostname AS_1_GY_2<br>!<br>! | 基本信息 |
| interface FastEthernet0/1<br>  switchport access vlan 110<br>  switchport mode access<br>!<br>interface FastEthernet0/2<br>  switchport access vlan 110<br>  switchport mode access<br>!<br>&lt;省略部分&gt;<br>interface FastEthernet0/16<br>  switchport access vlan 110 | F0/1-16 端口加入 VLAN 110 |
| interface FastEthernet0/17<br>!<br>&lt;省略部分&gt;<br>interface FastEthernet0/24<br>! | 其他端口默认配置，属于 VLAN 1 |
| interface Vlan1<br>  no ip address<br>  shutdown<br>!<br>&lt;省略部分&gt;<br>End | VLAN1 因为没有使用，出于安全考虑，应将其关闭 |

⑦ 其他设备参考 VLAN 规划表参数要求以及以上方法进行配置和验证,确认无误后,保存设备,并保存设备配置脚本文件。

⑧ 其他接入层设备配置脚本。

AS_1_GY_1 的脚本文件为 AS_1_GY_1_ Acc VL_CFG.txt,内容如下:

```
##################     as-1-gy-1-accvl-cfg     ##################
configure terminal
hostname AS_1_GY_1
 end
write
```

AS_1_GY_2 的脚本文件为 AS_1_GY_2_ Acc VL_CFG.txt,内容如下:

```
##################     as-1-gy-2-accvl-cfg     ##################
configure terminal
hostname AS_1_GY_2
vlan 110
 name 1_GY_2
 exit
interface range FastEthernet0/1 - 9
 switchport mode access
 switchport access vlan 130
 exit
interface FastEthernet0/24
 switchport mode trunk
 end
write
```

AS_2_GY_1 的脚本文件为 AS_2_GY_1_ Acc VL_CFG.txt,内容如下:

```
##################     as-2-gy-1-accvl-cfg     ##################
configure terminal
hostname AS_2_GY_1
vlan 120
 name 2_GY_1
 exit
interface range FastEthernet0/1 - 9
 switchport mode access
 switchport access vlan 130
 exit
interface FastEthernet0/24
 switchport mode trunk
 end
write
```

AS_2_GY_2 的脚本文件为 AS_2_GY_2_ Acc VL_CFG.txt,内容如下:

```
##################     as-2-gy-2-accvl-cfg     ##################
configure terminal
hostname AS_2_GY_2
vlan 130
 name 2_GY_2
 exit
```

```
interface range FastEthernet0/1 - 9
 switchport mode access
 switchport access vlan 130
 exit
interface FastEthernet0/24
 switchport mode trunk
 end
write
```

AS_3_GY_1 的脚本文件为 AS_3_GY_1_ Acc VL_CFG.txt，内容如下：

```
##################     as-3-gy-1-accvl-cfg     ##################
configure terminal
hostname  AS_3_GY_1
vlan 140
 name 3_GY_1
 exit
interface range FastEthernet0/1 - 9
 switchport mode access
 switchport access vlan 140
 exit
interface FastEthernet0/24
 switchport mode trunk
 end
write
```

AS_3_GY_2 的脚本文件为 AS_3_GY_2_ Acc VL_CFG.txt，内容如下：

```
##################     as-3-gy-2-accvl-cfg     ##################
configure terminal
hostname  AS_3_GY_2
vlan 150
 name 3_GY_2
 exit
interface range FastEthernet0/1 - 9
 switchport mode access
 switchport access vlan 150
 exit
interface FastEthernet0/24
 switchport mode trunk
 end
write
```

AS_4_GY_1 的脚本文件为 AS_4_GY_1_ Acc VL_CFG.txt，内容如下：

```
##################     as-4-gy-1-accvl-cfg     ##################
configure terminal
hostname  AS_4_GY_1
vlan 160
 name 4_GY_1
 exit
interface range FastEthernet0/1 - 9
 switchport mode access
```

```
 switchport access vlan 160
 exit
interface FastEthernet0/24
 switchport mode trunk
 end
write
```

AS_JX_1 的脚本文件为 AS_JX_1_ Acc VL_CFG.txt，内容如下：

```
##################     as-jx-1-accvl-cfg     ##################
configure terminal
hostname  AS_JX_1
vlan 200
 name JX_1
 exit
interface range FastEthernet0/1 - 9
 switchport mode access
 switchport access vlan 200
interface FastEthernet0/24
 switchport mode trunk
 end
write
```

AS_JX_2 的脚本文件为 AS_JX_2_ Acc VL_CFG.txt，内容如下：

```
##################     as-jx-2-accvl-cfg     ##################
configure terminal
hostname  AS_JX_2
vlan 210
 name JX_2
 exit
interface range FastEthernet0/1 - 9
 switchport mode access
 switchport access vlan 210
 exit
interface FastEthernet0/24
 switchport mode trunk
 end
write
```

AS_SX_1 的脚本文件为 AS_SX_1_ Acc VL_CFG.txt，内容如下：

```
##################     as-sx-1-accvl-cfg     ##################
configure terminal
hostname  AS_SX_1
vlan 220
 name SX_1
 exit
interface range FastEthernet0/1 - 9
 switchport mode access
 switchport access vlan 220
```

```
 exit
interface FastEthernet0/24
 switchport mode trunk
 end
write
```

AS_SX_2 的脚本文件为 AS_SX_2_ Acc VL_CFG.txt，内容如下：

```
##################     as-sx-2-accvl-cfg     ##################
configure terminal
hostname AS_SX_2
vlan 230
 name SX_2
 exit
interface range FastEthernet0/1 - 9
 switchport mode access
 switchport access vlan 230
 exit
interface FastEthernet0/24
 switchport mode trunk
 end
write
```

AS_TSG_1 的脚本文件为 AS_TSG_1_ Acc VL_CFG.txt，内容如下：

```
##################     as-tsg-1-accvl-cfg     ##################
configure terminal
hostname AS_TSG_1
vlan 240
 name Teach
 exit
interface range FastEthernet0/1 - 9
 switchport mode access
 switchport access vlan 240
 exit
interface FastEthernet0/24
 switchport mode trunk
 end
write
```

AS_TSG_2 的脚本文件为 AS_TSG_2_ Acc VL_CFG.txt，内容如下：

```
##################     as-tsg-2-accvl-cfg     ##################
configure terminal
hostname AS_TSG_2
vlan 250
 name Stu
 exit
interface range FastEthernet0/1 - 9
 switchport mode access
 switchport access vlan 250
 exit
```

```
interface FastEthernet0/24
 switchport mode trunk
 end
write
```

AS_ZH_1 的脚本文件为 AS_ZH_1_ Acc VL_CFG.txt，内容如下：

```
#################      as-zh-1-accvl-cfg      #################
configure terminal
hostname AS_ZH_1
vlan 260
 name Finan
 exit
interface range FastEthernet0/1 - 9
 switchport mode access
 switchport access vlan 260
 exit
interface FastEthernet0/24
 switchport mode trunk
 end
write
```

AS_ZH_2 的脚本文件为 AS_ZH_2_ Acc VL_CFG.txt，内容如下：

```
#################      as-zh-2-accvl-cfg      #################
configure terminal
hostname AS_ZH_2
vlan 270
 name Admin
 exit
interface range FastEthernet0/1 - 9
 switchport mode access
 switchport access vlan 270
 exit
interface FastEthernet0/24
 switchport mode trunk
 end
write
```

# 项目 4　部署汇聚 VLAN

## 项目描述

在项目 3 中，我们通过使用 VLAN 有效隔离部门间的广播，将端口划入特定 VLAN，如果交换机之间进行通信，使用 Access 模式无法实现多个 VLAN 在同一个端口（交换机级联）通信，如何控制特定的 VLAN 通过 Trunk（级联干道），是本项目需要分析解决的问题。

通过本项目，读者可以掌握以下技能：

① 能够配置 Trunk 端口。
② 能够分析 Trunk 封装过程。

③ 能够对 Trunk 流量进行优化。
④ 能够排除 Trunk 故障。

知识准备

接入层交换机上通常有多个 VLAN，通过上行链路与汇聚层交换机连接，汇聚层交换机的第一个作用就是联通部门 VLAN，我们首先分析一下以太网中数据的封装，了解 Trunk 的工作原理。

1）以太网帧的封装

以太网帧的格式如图 2-2 所示。

| 8 | 6 | 6 | 2 | 可变长 | 4 |
|---|---|---|---|---|---|
| 前导位 | 目标地址 | 源地址 | 类型 | 数据 | FCS |

图 2-2  以太网帧的格式

针对图 2-2 中各个字段解释如下：
① 前导位：表示新的帧开始。
② 目标地址：目标 MAC 地址。
③ 源地址：源 MAC 地址。
④ 类型：封装第三层数据类型，0800 表示 IP。
⑤ FCS：帧校验序列，用于检测帧传输过程中是否出错。

2）802.1q Trunk 封装格式

802.1q 的封装格式如图 2-3 所示。

| 8 | 6 | 6 | 4 | 2 | 可变长 | 4 |
|---|---|---|---|---|---|---|
| 前导位 | 目标地址 | 源地址 | 标签（tag） | 类型 | 数据 | FCS |

图 2-3  802.1q 封装格式

为区分不同 VLAN 信息，交换机会在原以太网帧的基础上插入 tag 字段（4 个字结，详见图 2-3），tag 字段的主要内容见图 2-4。

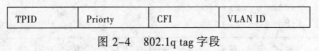

图 2-4  802.1q tag 字段

最常用的是基于端口的 VLAN，由此引出几种不同的端口类型。
① Access 端口：只能属于一个 VLAN，该端口接收到普通以太网帧就打上标签，VLAD ID 就是该端口所在 VLAN 的 ID，该端口向外转发时会将标签去掉，一般用于连接 PC 等设备。
② Trunk 端口：可以属于多个 VLAN，该端口转发帧时除了与其 PVID 一致的 VLAN 帧不打标签外，其他的均需打上标签；PVID 就是该端口的默认 VLAN 的 ID，一般用于交换机之间的互联。

注意

对于 CISCO 交换机 2950，默认使用 802.1q 封装协议，而 3560 以及更高设备需要先封装协议（ISL 或者 802.1q）才能形成 Trunk。

## 项目实施

### 1) 项目设计

将所有交换机之间的级联端口设置为 Trunk 模式，而不使用自动协商模式，Trunk 端口只允许特定的接入 VLAN 通过，以过滤不必要的广播流量。

> **注意**
> 项目讲解以 4#公寓楼为例，见图 2-5，其他进行类似配置即可。

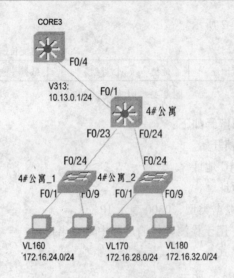

图 2-5　4#公寓汇聚部署

### 2) 项目实施

① 在汇聚层交换机 DS_GY_4 上使用 swtichport mode trunk 命令配置下行端口 fastethernet 0/24 为 trunk：

```
DS_GY_4 (config)#interface range fastethernet 0/24
DS_GY_4 (config-range-if)#switchport trunk encapsulation dot1q
DS_GY_4 (config-range-if)#switchport mode trunk
DS_GY_4 (config-range-if)#exit
```

② 使用 show vlan brief 命令查看 VLAN 与端口信息，发现 F0/24 口不属于任何 VLAN，有别于 Access 模式：

```
DS_4_GY#show vlan brief
VLAN Name                             Status    Ports
---- -------------------------------- --------- -------------------------------
1    default                          active    Fa0/2, Fa0/3, Fa0/4, Fa0/5
                                                Fa0/6, Fa0/7, Fa0/8, Fa0/9
                                                Fa0/10, Fa0/11, Fa0/12, Fa0/13
                                                Fa0/14, Fa0/15, Fa0/16, Fa0/17
                                                Fa0/18, Fa0/19, Fa0/20, Fa0/21
                                                Fa0/22, Gig0/1, Gig0/2
```

```
160  4_GY_1                    active      Fa0/23
170  4_GY_2_1                  active
180  4_GY_2_2                  active
<省略部分>
```

③ 使用 show vlan brief 命令查看 VLAN 与端口信息，发现 F0/24 口不属于任何 VLAN，有别于 Access 模式。

④ 使用 show interfaces fastethernet 0/24 switchport 命令查看 VLAN 与端口信息：

```
DS_4_GY#show interfaces fastethernet 0/24 switchport
Name: Fa0/24
Switchport: Enabled
Administrative Mode: trunk
Operational Mode: trunk
Administrative Trunking Encapsulation: dot1q
Operational Trunking Encapsulation: dot1q
Negotiation of Trunking: On
<省略部分>
```

可以看到端口的 Administrative Mode、Operational Mode 默认均为 trunk，封装协议为 dot1q。

⑤ 在下行设备 AS_4_GY_2 上使用 show interfaces fastethernet 0/24 switchport 命令查看 VLAN 与端口信息：

```
AS_4_GY_2#show interfaces fastethernet 0/24 switchport
Name: Fa0/24
Switchport: Enabled
Administrative Mode: dynamic auto
Operational Mode: trunk
Administrative Trunking Encapsulation: dot1q
Operational Trunking Encapsulation: dot1q
Negotiation of Trunking: On
<省略部分>
```

AS_4_GY_2 的 F0/24 口为默认设置，可以看到工作在 trunk 模式，封装协议为 dot1q，因为端口的默认管理模式为 dynamic auto，如果两个交换机使用端口 1 和端口 2 进行连接，会根据对端设备协商是否形成 trunk。能否形成 trunk，可见表 2-6。

表 2-6  管理模式与 trunk

| 端口 1 管理模式 | 端口 2 管理模式 | 能否形成 trunk |
| --- | --- | --- |
| dynamic auto | dynamic auto | 否 |
| dynamic auto | trunk | 是 |
| dynamic desirable | dynamic auto | 是 |
| dynamic desirable | trunk | 是 |
| trunk | trunk | 是 |

⑥ 在 DS_GY_4 上，创建 VLAN170、VLAN180，配置 VLAN 170、VLAN180 的 IP 地址为 172.16.28.254、172.16.32.254，子网掩码均为 255.255.255.0。

⑦ 配置主机 PC4、PC5 地址如表 2-7 所示。

表 2-7  主机地址信息表

| 主　机 | IP 地址 | 子网掩码 | 默认网关 | MAC 地址 |
|---|---|---|---|---|
| PC4 | 172.16.28.1 | 255.255.255.0 | 172.16.28.254 | 00E0.F911.B84A |
| PC5 | 172.16.32.1 | 255.255.255.0 | 172.16.32.254 | 0002.16EB.C7D3 |

⑧ 在 PC4 上 ping PC5，见图 2-6，在 Packet Tracer 中进行抓包分析。

a. 当数据帧由 F0/1 口进入接入层交换机时，L2 封装为标准以太网帧，见图 2-7。

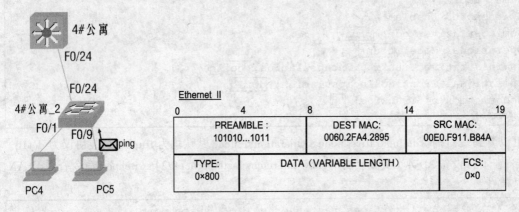

图 2-6  PC4 ping PC5　　　　　　　　　图 2-7  以太网帧

b. 当数据帧由 F0/24 口离开接入层交换机进入 Trunk 链路时，L2 封装为 802.1q 格式（TCI 字段 0xaa 即为 VLAN ID：170），见图 2-8。

图 2-8  802.1q 帧

c. 当数据帧进入汇聚层交换机时，L2 封装为 802.1q 格式（TCI 字段 0xaa 即为 VLAN 170）。

d. 当数据帧离开汇聚层交换机时，L2 封装为 802.1q 格式（TCI 字段 0xb4 即为 VLAN ID：180），帧转发到 VLAN 180，见图 2-9。

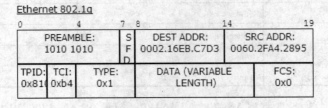

图 2-9  802.1q 帧

e. 当数据帧再次进入接入层交换机 Trunk 端口时，L2 封装为 802.1q 格式（TCI 字段 0xb4 即为 VLAN ID：180）。

f. 当数据帧再次进入接入层交换机 Trunk 端口时，L2 将去掉标签部分，成为以太网帧，见图 2-10。

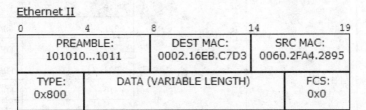

图 2-10　以太网帧

3）总结

通过抓包分析，以太网帧传输到 Trunk 链路，将封装特定的 VLAN 标签，以确定该 VLAN 能否通过，当离开 Trunk 链路时，将去掉 tag 标签。

4）Trunk 流量过滤

对于 Trunk 端口，默认情况下，允许所有 VLAN 信息通过，可以使用 switchport trunkallow 命令进行 VLAN 处理，详细配置命令可以参考表 2-8。

表 2-8　Trunk 配置命名

| 序　号 | 命　令 | 示　例 | 说　明 |
|---|---|---|---|
| 1 | word | 10 | 只允许 VLAN 10 通过 |
| 2 | add | add 10 | 在原有 VLAN 列表上增加 VLAN 10 |
| 3 | all | all | 允许所有 VLAN 通过，默认设置 |
| 4 | except | except 10 | 允许除 VLAN 10 之外的所有 VLAN 通过 |
| 5 | none | none | 不允许任何 VLAN |
| 6 | remove | remove 10 | 从当前允许列表中移除 VLAN 10 |

详细命令可参考思科产品配置手册。

① 汇聚 VLAN 与接入 VLAN 基本类似，汇聚 VLAN 是连接不同层次的交换机（通常为三层交换机），而接入 VLAN 更多的是连接终端。

② 汇聚 VLAN 设计，详细见表 2-9。

表 2-9　汇聚 VLAN 设计

| 楼　宇 | 部　门 | 设　备 | VLAN ID | 端　口 |
|---|---|---|---|---|
| 1#公寓 | 1#公寓 | DS_1_GY | V310 | F0/1 |
| 2#公寓 | 2#公寓 | DS_2_GY | V311 | F0/1 |
| 3#公寓 | 3#公寓 | DS_3_GY | V312 | F0/1 |
| 4#公寓 | 4#公寓 | DS_4_GY | V313 | F0/1 |
| 教学楼 | 教学楼 | DS_JX | V320 | F0/1 |
| 实训楼 | 实训楼 | DS_SX | V321 | F0/1 |

续表

| 楼　宇 | 部　门 | 设　备 | VLAN ID | 端　口 |
|---|---|---|---|---|
| 图书馆 | 图书馆 | DS_TSG | V322 | F0/1 |
| 综合楼 | 综合楼 | DS_ZH | V323 | F0/1 |
| 信息中心 | 信息中心 | CORE2 | V320 | F0/1 |
| | | | V321 | F0/1 |
| | | | V322 | F0/1 |
| | | | V323 | F0/1 |
| | | CORE3 | V310 | F0/1 |
| | | | V311 | F0/1 |
| | | | V312 | F0/1 |
| | | | V313 | F0/1 |

③ 为联通不同的 VLAN，在汇聚层交换机上配置 IP 地址，作为相应 VLAN 内主机的网关，能实现同一个汇聚层交换机连接的 VLAN 内主机的通信。

④ 如果让多个汇聚层交换机所连接的 VLAN 能相互 ping 通，读者可以使用静态路由或者动态路由进行，关于路由的内容，将在下一章详细讨论。

### 工程化操作

① 安装网络设备，保证物理联通。

② 创建配置脚本：本项目先配置 CORE3—DS_4_GY—AS_4_GY_2 三层设备，见图 2-11。新建文本文件，将各个设备配置命令复制、粘贴到其中，保存文件名为：设备名_ Trk VL_CFG.txt。

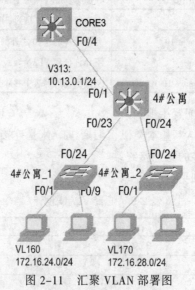

图 2-11　汇聚 VLAN 部署图

创建脚本文件为 CORE3_ Trk VL_CFG.txt，内容如下：

```
################## core3-trk vl-cfg ##################
configure terminal
```

```
hostname core3
vlan 302
 name Core13
 exit
vlan 303
 name Core23
 exit
vlan 310
 name GY1
 exit
vlan 311
 name GY2
 exit
vlan 312
 name GY3
 exit
vlan 313
 name GY4
 exit
interface FastEthernet0/1
 switchport mode access
 switchport access vlan 310
 exit
interface FastEthernet0/2
 switchport mode access
 switchport access vlan 311
 exit
interface FastEthernet0/3
 switchport mode access
 switchport access vlan 312
 exit
interface FastEthernet0/4
 switchport mode access
 switchport access vlan 313
 exit
interface FastEthernet0/23
 switchport mode access
 switchport access vlan 302
 exit
interface FastEthernet0/24
 switchport mode access
 switchport access vlan 303
 exit
<省略部分>
```

创建脚本文件为 DS_4_GY _ Trk VL_CFG.txt，内容如下：

```
##################    ds-4-gy- trk vl-cfg    ##################
configure terminal
hostname DS_4_GY
vlan 313
```

```
 name GY4
 exit
vlan 160
 name 4_GY_1
 exit
vlan 170
 name 4_GY_2_1
 exit
vlan 180
 name 4_GY_2_2
 exit
interface FastEthernet0/1
 switchport mode access
 switchport access vlan 313
 exit
interface FastEthernet0/23
 switchport mode access
 switchport access vlan 160
 exit
interface FastEthernet0/24
 switchport trunk encapsulation dot1q
 switchport mode trunk
 exit
interface vlan 313
 no shutdown
 ip address 10.13.0.254 255.255.255.0
 exit
interface vlan 160
 no shutdown
 ip address 172.16.24.254 255.255.255.0
 exit
interface vlan 170
 no shutdown
 ip address 172.16.28.254 255.255.255.0
 exit
interface vlan 180
 no shutdown
 ip address 172.16.32.254 255.255.255.0
 exit
<省略部分>
```

创建脚本文件为 AS_4_GY_2_ Trk VL_CFG.txt，内容如下：

```
################   as-4-gy-2-trk vl-cfg   ################
configure terminal
hostname AS_4_GY_2
vlan 170
 name 4_GY_2_1
```

```
  exit
vlan 180
 name 4_GY_2_2
 exit
interface range FastEthernet0/1 - 8
 switchport mode access
 switchport access vlan 170
 exit
interface range FastEthernet0/9 - 16
 switchport mode access
 switchport access vlan 180
 exit
interface FastEthernet0/24
 switchport mode trunk
 exit
<省略部分>
```

实际工程项目中,两个交换机连接形成 Trunk,建议在每个端口手工封装并配置为 Trunk 模式。

③ 执行配置脚本:使用 CONSOLE 端口或者远程连接登录设备,打开超级终端或者 SecureCRT 程序,进入特权模式,复制设备中的命令脚本(设备名_CFG.txt),在超级终端或者 SecureCRT 程序中右击选择"粘贴",执行脚本。

④ 验证配置:在各设备的特权模式下,使用 show running-config 命令检查当前运行配置,注意分析配置中的各部分内容。

CORE3 运行配置如表 2-10 所示。

表 2-10  CORE3 运行配置

| 配 置 | 备 注 |
| --- | --- |
| core3#show running-config | 查看运行配置 |
| <省略部分><br>interface FastEthernet0/1<br> switchport access vlan 310<br> switchport mode access<br>!<br>interface FastEthernet0/2<br> switchport access vlan 311<br> switchport mode access<br>!<br>interface FastEthernet0/3<br> switchport access vlan 312<br> switchport mode access<br>!<br>interface FastEthernet0/4<br> switchport access vlan 313<br> switchport mode access<br>! | 连接核心各公寓楼汇聚上行端口 |

续表

| 配 置 | 备 注 |
|---|---|
| <省略部分><br>interface FastEthernet0/23<br>  switchport access vlan 302<br>  switchport mode access<br>!<br>interface FastEthernet0/24<br>  switchport access vlan 303<br>  switchport mode access<br>! | 连接核心 CORE1、CORE2 的 VLAN 与接口 |
| interface Vlan302<br>  ip address 10.8.2.254 255.255.255.0<br>!<br>interface Vlan303<br>  ip address 10.8.3.254 255.255.255.0<br>! | 配置 VLAN 的 IP 地址，与核心 CORE1、CORE2 连接 |
| interface Vlan310<br>  ip address 10.10.0.1 255.255.255.0<br>!<br>interface Vlan311<br>  ip address 10.11.0.1 255.255.255.0<br>!<br>interface Vlan312<br>  ip address 10.12.0.1 255.255.255.0<br>!<br>interface Vlan313<br>  ip address 10.13.0.1 255.255.255.0<br>!<br><省略部分> | 配置 VLAN IP 地址，与公寓楼汇聚交换机连接 |

DS_4_GY 运行配置如表 2-11 所示。

表 2-11　DS-4-GY 运行配置

| 配 置 | 备 注 |
|---|---|
| DS_4_GY#show running-config | 查看运行配置 |
| <省略部分><br>hostname DS_4_GY<br>!<br>interface FastEthernet0/1<br>  switchport access vlan 313<br>  switchport mode access<br>! | 连接核心 CORE3 端口 |

续表

| 配　　　置 | 备　　　注 |
|---|---|
| <省略部分><br>interface FastEthernet0/23<br>　switchport access vlan 160<br>　switchport mode access<br>! | 因为 F0/23 口所连接接入层交换机只有 VLAN160，所以未配置为 Trunk，可以配置类似于 F0/24 |
| interface FastEthernet0/24<br>　switchport trunk encapsulation dot1q<br>　switchport mode trunk<br>!<br><省略部分> | 配置 F0/24 口为 Trunk 模式 |
| interface Vlan160<br>　ip address 172.16.24.254 255.255.255.0<br>!<br>interface Vlan170<br>　ip address 172.16.28.254 255.255.255.0<br>!<br>interface Vlan180<br>　ip address 172.16.32.254 255.255.255.0<br>! | 配置 VLAN 的 IP 地址，作为响应主机的网关地址 |
| interface Vlan313<br>　ip address 10.13.0.254 255.255.255.0<br>!<br><省略部分> | 配置 VLAN 的 IP 地址，与核心 CORE3 连接 |

- AS_4_GY_2 运行配置如表 2-12 所示。

表 2-12　AS-4-GY-2 运行配置

| 配　　　置 | 备　　　注 |
|---|---|
| AS_4_GY_2#show running-config | 查看运行配置 |
| <省略部分><br>interface FastEthernet0/1<br>　switchport access vlan 170<br>　switchport mode access<br>!<br>interface FastEthernet0/2<br>　switchport access vlan 170<br>　switchport mode access<br>!<br>interface FastEthernet0/3<br>　switchport access vlan 170<br>　switchport mode access<br>!<br>interface FastEthernet0/4<br>　switchport access vlan 170 | VLAN 170 内端口 |

续表

| 配置 | 备注 |
|---|---|
| switchport mode access<br>!<br>interface FastEthernet0/5<br>　switchport access vlan 170<br>　switchport mode access<br>!<br>interface FastEthernet0/6<br>　switchport access vlan 170<br>　switchport mode access<br>!<br>interface FastEthernet0/7<br>　switchport access vlan 170<br>　switchport mode access<br>!<br>interface FastEthernet0/8<br>　switchport access vlan 170<br>　switchport mode access<br>! | VLAN 170 内端口 |
| interface FastEthernet0/9<br>　switchport access vlan 180<br>　switchport mode access<br>!<br>interface FastEthernet0/10<br>　switchport access vlan 180<br>　switchport mode access<br>!<br>interface FastEthernet0/11<br>　switchport access vlan 180<br>　switchport mode access<br>!<br>interface FastEthernet0/12<br>　switchport access vlan 180<br>　switchport mode access<br>!<br>interface FastEthernet0/13<br>　switchport access vlan 180<br>　switchport mode access<br>!<br>interface FastEthernet0/14<br>　switchport access vlan 180<br>　switchport mode access<br>! | VLAN 180 内端口 |

续表

| 配 置 | 备 注 |
|---|---|
| interface FastEthernet0/15<br>　switchport access vlan 180<br>　switchport mode access<br>!<br>interface FastEthernet0/16<br>　switchport access vlan 180<br>　switchport mode access<br>!<br><省略部分> | VLAN 180 内端口 |
| interface FastEthernet0/24<br>　switchport mode trunk<br>!<br><省略部分> | 配置上行端口为 Trunk |

⑤ 其他设备参考 VLAN 规划表、图 1-3 及以上方法进行配置和验证，确认无误后，保存设备，并保存设备配置脚本文件。

⑥ 核心层设备配置脚本。

CORE_2 的脚本文件为 CORE_2_ Trk VL _CFG.txt，内容如下：

```
################## core-2-trk vl-cfg ##################
configure terminal
hostname CORE_2
vlan 320
 name Teaching
 exit
vlan 321
 name Pracising
 exit
vlan 322
 name lib
 exit
vlan 323
 name common
 exit
interface FastEthernet0/1
 switchport mode access
 switchport access vlan 320
 exit
interface FastEthernet0/2
 switchport mode access
 switchport access vlan 321
 exit
interface FastEthernet0/3
 switchport mode access
 switchport access vlan 322
```

```
 exit
interface FastEthernet0/4
 switchport mode access
 switchport access vlan 323
 exit
interface vlan 320
 no shutdown
 ip address 10.20.0.1 255.255.255.0
 exit
interface vlan 321
 no shutdown
 ip address 10.21.0.1 255.255.255.0
 exit
interface vlan 322
 no shutdown
 ip address 10.22.0.1 255.255.255.0
 exit
interface vlan 323
 no shutdown
 ip address 10.23.0.1 255.255.255.0
 end
write
```

CORE_3 的脚本文件为 CORE_3_ Trk VL _CFG.txt，内容如下：

```
##################    core-3-trk vl-cfg    ##################
configure terminal
hostname  CORE_3
vlan 310
 name GY1
 exit
vlan 311
 name GY2
 exit
vlan 312
 name GY3
 exit
vlan 313
 name GY4
 exit
interface FastEthernet0/1
 switchport mode access
 switchport access vlan 310
 exit
interface FastEthernet0/2
 switchport mode access
 switchport access vlan 311
 exit
interface FastEthernet0/3
 switchport mode access
 switchport access vlan 312
```

```
 exit
interface FastEthernet0/4
 switchport mode access
 switchport access vlan 313
 exit
interface FastEthernet0/23
 switchport mode access
 switchport access vlan 302
 exit
interface vlan 310
 no shutdown
 ip address 10.10.0.1 255.255.255.0
 exit
interface vlan 311
 no shutdown
 ip address 10.11.0.1 255.255.255.0
 exit
interface vlan 312
 no shutdown
 ip address 10.12.0.1 255.255.255.0
 exit
interface vlan 313
 no shutdown
 ip address 10.13.0.1 255.255.255.0
 end
write
```

⑦ 汇聚层设备配置脚本。

DS_1_GY 的脚本文件为 DS_1_GY_ Trk VL _CFG.txt，内容如下：

```
##################     ds-1-gy-trk vl-cfg     ##################
configure terminal
hostname  DS_1_GY
vlan 310
 name GY1
 exit
interface FastEthernet0/1
 switchport mode access
 switchport access vlan 310
 exit
interface vlan 310
 no shutdown
 ip address 10.10.0.254 255.255.255.0
 end
write
```

DS_2_GY 的脚本文件为 DS_2_GY_ Trk VL _CFG.txt，内容如下：

```
##################     ds-2-gy-trk vl-cfg     ##################
configure terminal
hostname  DS_2_GY
vlan 311
```

```
 name GY2
 exit
interface FastEthernet0/1
 switchport mode access
 switchport access vlan 311
 exit
interface vlan 311
 no shutdown
 ip address 10.11.0.254 255.255.255.0
 end
write
```

DS_3_GY 的脚本文件为 DS_3_GY_ Trk VL _CFG.txt,内容如下:

```
##################     ds-3-gy-trk vl-cfg     ##################
configure terminal
hostname DS_3_GY
vlan 312
 name GY3
 exit
interface FastEthernet0/1
 switchport mode access
 switchport access vlan 312
 exit
interface vlan 312
 no shutdown
 ip address 10.12.0.254 255.255.255.0
 end
write
```

DS_JX 的脚本文件为 DS_JX_ Trk VL _CFG.txt,内容如下:

```
##################     ds-jx-trk vl-cfg     ##################
configure terminal
hostname DS_JX
vlan 320
 name Teaching
 exit
interface FastEthernet0/1
 switchport mode access
 switchport access vlan 320
 exit
interface vlan 320
 no shutdown
 ip address 10.20.0.254 255.255.255.0
 end
write
```

DS_SX 的脚本文件为 DS_SX_ Trk VL _CFG.txt,内容如下:

```
##################     ds-sx-trk vl-cfg     ##################
configure terminal
hostname DS_SX
vlan 321
```

```
 name Practising
 exit
interface FastEthernet0/1
 switchport mode access
 switchport access vlan 321
 exit
interface vlan 321
 no shutdown
 ip address 10.21.0.254 255.255.255.0
 end
write
```

DS_TSG 的脚本文件为 DS_TSG_ Trk VL _CFG.txt，内容如下：

```
##################     ds-tsg-trk vl-cfg     ##################
configure terminal
hostname  DS_TSG
vlan 322
 name Lib
 exit
interface FastEthernet0/1
 switchport mode access
 switchport access vlan 322
 exit
interface vlan 322
 no shutdown
 ip address 10.22.0.254 255.255.255.0
 end
write
```

DS_ZH 的脚本文件为 DS_ZH Trk VL _CFG.txt，内容如下：

```
##################     ds-zh-trk vl-cfg     ##################
configure terminal
hostname  DS_ZH
vlan 323
 name common
 exit
interface FastEthernet0/1
 switchport mode access
 switchport access vlan 323
 exit
interface vlan 323
 no shutdown
 ip address 10.23.0.254 255.255.255.0
end
```

# 项目 5　部署管理 VLAN

## 项目描述

在项目 4 中，在路由联通的基础上，我们可以在任何一台 PC 上 Telnet 到校园网中任意一个设备，但是如果路由出现故障，如何保证网络管理员能远程排错呢？下面将重点讨论管理 VLAN 的作用域部署。

通过本项目，读者可以掌握以下技能：

① 能够配置 Native VLAN。

② 能够分析管理 VLAN 的作用。

③ 能够设计部署管理 VLAN。

知识准备

1）管理 VLAN

在设备连接之后，通常远程进行配置，使用管理 VLAN 可以在路由不同的情况下，在 OSI L2 层进行管理。默认情况下，交换机的管理 VLAN 均为 1，为安全起见，在项目中创建 VLAN 99 作为管理 VLAN，同时作为 Native VLAN 进行配置。

2）Native VLAN

Native VLAN 即本征 VLAN，和其他 VLAN 的区别在于：非 Native VLAN 在 Trunk 中传输数据时要被打上标签（如 dot1q 或者 isl），但是 Native VLAN 在 Trunk 中传输数据时是不打标签的。

所有的帧在 Trunk 中都会打上标签，不同点在于，如果帧在进入 Trunk 以前已经打上标签了，比如 VLAN 2 的标签，并且 Trunk 允许 VLAN 2 通过的话，该 VLAN 2 的帧就通过，反之丢弃。另外，如果帧在进入 Trunk 时没有被打上标签，那么 Trunk 就会给它打上 Native VLAN 的标签，该帧在 Trunk 中就以 Native VLAN 的身份传输，Native VLAN 是用于 Trunk 口的，在 Access 口没有 Native VLAN 的概念。在一些协议中，如 STP，交换机之间是要互相协商通信的，如果对 STP 的数据包打了标签的话，会导致一些不支持 VLAN 交换机不能相互协商。为了解决这个问题，提出 Native VLAN 的概念。在 Trunk 中，对于不带标签的流入数据，在交换机中设置 Native VLAN ID，流出时，当发现标签是该端口的 Native VLAN ID，去掉标签转发。

项目实施

1）项目设计

将所有交换机之间的级联端口设置为 Trunk 模式，而不使用自动协商模式，Trunk 口只允许特定的接入 VLAN 通过，所有交换机都配置 Native VLAN 为 99，配置相同网段 IP 地址，能实现二层管理。

2）项目实施

将交换机之间的级联链路设置为 Trunk，为减少广播流量，限制 Trunk 链路上允许通过的 VLAN 信息，配置管理 VLAN 和 Native VLAN 为 99。

> 注意
>
> 项目讲解以 4#公寓楼为例，其他类似进行配置即可。

① 在各个设备上创建管理 VLAN，将 Native VLAN 和管理 VLAN 配置成相同，并配置管理地址，详见表 2-13。

表 2-13 管理 VLAN 设计表

| 楼宇 | 部门 | 设　　备 | VLAN ID | VLAN NAME | IP | 掩　　码 |
|---|---|---|---|---|---|---|
| 1#公寓 | 1#公寓 | DS_1_GY | V99 | manage | 10.0.0.10 | 255.255.255.0 |
| | | AS_1_GY_1 | V99 | manage | 10.0.0.11 | 255.255.255.0 |
| | | AS_1_GY_2 | V99 | manage | 10.0.0.12 | 255.255.255.0 |
| 2#公寓 | 2#公寓 | DS_2_GY | V99 | manage | 10.0.0.20 | 255.255.255.0 |
| | | AS_2_GY_1 | V99 | manage | 10.0.0.21 | 255.255.255.0 |
| | | AS_2_GY_2 | V99 | manage | 10.0.0.22 | 255.255.255.0 |
| 3#公寓 | 3#公寓 | DS_3_GY | V99 | manage | 10.0.0.30 | 255.255.255.0 |
| | | AS_3_GY_1 | V99 | manage | 10.0.0.31 | 255.255.255.0 |
| | | AS_3_GY_2 | V99 | manage | 10.0.0.32 | 255.255.255.0 |
| 4#公寓 | 4#公寓 | DS_4_GY | V99 | manage | 10.0.0.40 | 255.255.255.0 |
| | | AS_4_GY_1 | V99 | manage | 10.0.0.41 | 255.255.255.0 |
| | | AS_4_GY_2 | V99 | manage | 10.0.0.42 | 255.255.255.0 |
| 教学楼 | 教学楼 | DS_JX | V99 | manage | 10.0.0.50 | 255.255.255.0 |
| | | AS_JX_1 | V99 | manage | 10.0.0.51 | 255.255.255.0 |
| | | AS_JX_2 | V99 | manage | 10.0.0.52 | 255.255.255.0 |
| 实训楼 | 实训楼 | DS_SX | V99 | manage | 10.0.0.60 | 255.255.255.0 |
| | | AS_SX_1 | V99 | manage | 10.0.0.61 | 255.255.255.0 |
| | | AS_SX_2 | V99 | manage | 10.0.0.62 | 255.255.255.0 |
| 图书馆 | 图书馆 | DS_TSG | V99 | manage | 10.0.0.70 | 255.255.255.0 |
| | | AS_TSG_1 | V99 | manage | 10.0.0.71 | 255.255.255.0 |
| | | AS_TSG_2 | V99 | manage | 10.0.0.72 | 255.255.255.0 |
| 综合楼 | 综合楼 | DS_ZH | V99 | manage | 10.0.0.80 | 255.255.255.0 |
| | | AS_ZH_1 | V99 | manage | 10.0.0.81 | 255.255.255.0 |
| | | AS_ZH_2 | V99 | manage | 10.0.0.82 | 255.255.255.0 |
| 主楼（信息中心） | 主楼（信息中心） | CORE1 | V99 | manage | 10.0.0.91 | 255.255.255.0 |
| | | CORE2 | V99 | manage | 10.0.0.92 | 255.255.255.0 |
| | | CORE3 | V99 | manage | 10.0.0.93 | 255.255.255.0 |
| | | CORE4 | V99 | manage | 10.0.0.94 | 255.255.255.0 |

② 在 AS_4_GY_2 交换机上将 F0/1 口划入 VLAN 99，配置 PC4 的 IP 地址为 10.0.0.100，掩码为 255.255.255.0。

③ 在 PC4 上尝试 Telnet 到 CORE3、DS_4_GY、DS_4_GY_2 交换机。

## 工程化操作

① 安装网络设备，保证物理联通。

② 创建配置脚本：本项目先配置 CORE3—DS_4_GY—AS_4_GY_2 三层设备。新建文本文件，将各个设备配置命令复制、粘贴到其中，保存文件名为：设备名_ Mng VL_CFG.txt。

创建脚本文件为 CORE3_Mng VL_CFG.txt，内容如下：

```
##################     core3- mng vl-cfg     ##################
configure terminal
enable secure cisco
  line vty 0 4
  login
  pass cisco
  exit
vlan 99
  name manage
  exit
interface vlan 99
  no shutdown
  Ip address 10.0.0.93 255.255.255.0
  exit
interface fastethernet 0/1
  switchport trunk encapsulation dot1q
  switchport mode trunk
  switchport trunk native vlan 99
  end
interface fastethernet 0/2
  switchport trunk encapsulation dot1q
  switchport mode trunk
  switchport trunk native vlan 99
interface fastethernet 0/3
  switchport trunk encapsulation dot1q
  switchport mode trunk
  switchport trunk native vlan 99
  end
interface fastethernet 0/4
  switchport trunk encapsulation dot1q
  switchport mode trunk
  switchport trunk native vlan 99
interface fastethernet 0/23
  switchport trunk encapsulation dot1q
  switchport mode trunk
  switchport trunk native vlan 99
  end
interface fastethernet 0/24
  switchport trunk encapsulation dot1q
  switchport mode trunk
  switchport trunk native vlan 99
```

创建脚本文件为 DS_4_GY _Mng VL_CFG.txt，内容如下：

```
##################     ds-4-gy- mng vl-cfg     ##################
configure terminal
enable secure cisco
line vty 0 4
  login
  pass cisco
```

```
  exit
vlan 99
  name manage
  exit
interface vlan 99
  no shutdown
  ip address 10.0.0.40 255.255.255.0
  exit
interface fastethernet 0/1
  switchport trunk encapsulation dot1q
  switchport mode trunk
  Switchport trunk native vlan 99
  exit
interface fastethernet 0/23
  switchport trunk encapsulation dot1q
  switchport mode trunk
  switchport trunk native vlan 99
  exit
interface fastethernet 0/24
  switchport trunk encapsulation dot1q
  switchport mode trunk
  switchport trunk native vlan 99
```

创建脚本文件为 AS_4_GY_2 _Mng VL_CFG.txt，内容如下：

```
##################    as-4-gy-2- mng vl-cfg    ##################
configure terminal
enable secure cisco
line vty 0 4
  login
  pass cisco
  exit
vlan 99
  name manage
  exit
Interface vlan 99
  no shutdown
  ip address 10.0.0.42 255.255.255.0
  exit
interface fastethernet 0/24
  switchport mode trunk
  switchport trunk native vlan 99
  switchport trunk allow vlan add 99
```

③ 执行配置脚本：使用 CONSOLE 端口或者远程连接登录设备，打开超级终端或者 SecureCRT 程序，进入特权模式，复制设备中的命令脚本，在超级终端或者 SecureCRT 程序中右击选择"粘贴"，执行脚本。

④ 验证配置：在各设备的特权模式下，使用 show running-config 命令检查当前运行配置；使用 show vlan 命令检查 VLAN；使用 show interface fastethernet 0/* trunk 命令检查 Trunk 端口配置。

⑤ 其他设备参考 VLAN 规划表以及以上方法进行配置和验证，确认无误后，保存设备，并保存设备配置脚本文件。

## 本章训练内容

1. 完成第 8 章项目 17 中任务 2。
2. 完成项目 5 中所有管理 VLAN 的部署。

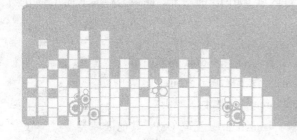

# 第 3 章 校园网内网优化与管理

高校校园网的用户通常以万计，通过 VLAN 设计，会有数十甚至数百个网络或者子网，而且设备数量庞大，通过前面两个章节的静态路由配置，相信大家体会很深，静态路由在配置大型校园网时非常烦琐，而且当网络链路故障或者网络升级改造时，静态路由调整和添加工作量比较大，稍不细心可能会导致路由无法联通，很明显，静态路由不是大型校园网的路由部署最佳选择。

在路由联通的基础上，部署校园网的访问控制策略，限制某些部门之间的访问，部署内网的安全等是校园网需要解决的重点问题。

通过本章项目的实践，可以学会校园网 OSPF 的区域规划、部署，通过配置访问控制列表（ACL）和安全，提高网络的效率、安全性。

本章需要完成的项目有：

项目 6——优化 OSPF 路由；

项目 7——访问控制部署；

项目 8——内网安全部署。

## 项目 6　优化 OSPF 路由

### 项目描述

随着校园网信息化建设的发展，校园网的用户规模已超万人，要求使用 OSPF 路由联通全网，同时通过对 OSPF 的区域设计优化路由流量。

通过本项目，读者可以掌握以下技能：

① 能够配置 OSPF。

② 能够分析 LSA 的类型。

③ 能够对 OSPF 区域进行分析。

④ 能够对 stub、nssa 等进行合理部署。

### 知识准备

动态路由协议可以解决静态路由的相关问题，典型的 IGP 路由协议有：RIP、EIGRP、OSPF。RIP 由于存在最大跳数 15 的限制，注定 RIP 不适用于大型网络。EIGRP 是一种很好的路由协议，具有很多优良特性，但是最大的问题是，它是思科私有的协议，必须在全网均为思科设备的网络上才能运行。在校园网中通常会有不同品牌的厂商设备，所以 OSPF 路由协议作为标准路由协议，更加适合园区网络的部署，而且 OSPF 通过区域（area）的划分，能很大程度上优化路由流量。

1）区域（area）与路由器角色

OSPF 的网络设计要求是双层层次化（2-layer hierarchy），包括如下两层：
- transit area（传输区域），也称 backbone area（骨干区域）或 area 0。
- regular area（规则区域），也称 nonbackbone area（非骨干区域）。

transit area 主要负责 IP 包快速和有效地传输；transit area 互联 OSPF 其他区域类型。一般来说，这个区域里不会出现端用户（end user）。

regular area 主要负责连接用户和资源；regular area 一般是根据功能和地理位置来划分的。一般一个 regular area 不允许其他区域的流量通过它到达另外一个区域，必须穿越 transit area。regular area 还可以有很多子类型，比如 stub area、totally stub area 和 not-so-stubby area。

在 OSPF 协议中，所有的路由器都保持有 LSDB，OSPF 路由器越多，LSDB 就越大，这可能对了解完整的网络信息有帮助，但是随着网络规模的扩大，可扩展性的问题就会越来越严重，采用的折中方案就是引入区域的概念。在某一个区域里的路由器只保持有该区域中所有路由器或链路的详细信息和其他区域的一般信息。当某个路由器或某条链路出现故障以后，信息只会在那个区域以内在邻居之间传递，那个区域以外的路由器不会收到信息。OSPF 要求层次化的网络设计，意味着所有的区域都要和 area 0 直接相连，如图 3-1 所示。

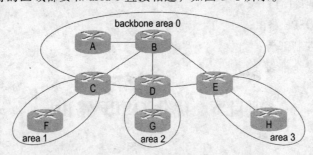

图 3-1　OSPF 区域

area 1、area 2 和 area 3 之间的连接是不允许的，它们都必须通过 backbone area 0 进行连接。建议每个区域中路由器的数量为 50 到 100 个，构建 area 0 的路由器称为骨干路由器（backbone router，BR），图 3-1 中 A 和 B 就是 BR；区域边界路由器（area border router，ABR）连接 area 0 和 nonbackbone area，图 3-1 中 C、D 和 E 就是 ABR。ABR 通常具有以下特征：
- 分隔 LSA 洪泛的区域。
- 是区域地址汇总的主要因素。
- 一般作为默认路由的源头。
- 为每个区域保持 LSDB。

理想的设计是使每个 ABR 只连接两个区域：backbone area 和其他区域，3 个区域为上限。

2）OSPF 邻接（adjacency）

运行 OSPF 的路由器通过交换 hello 包和别的路由器建立邻接（adjacency）关系，过程如下：

① 路由器和别的路由器交换 hello 包，目标地址采用多播地址。

② hello 包交换完毕，邻接关系形成。

③ 通过交换 LSA 和对接收方的确认进行同步 LSDB。对于 OSPF 路由器而言，进入完全邻接状态。

④ 如果需要的话，路由器转发新的 LSA 给其他的邻居，来保证整个区域内 LSDB 的完全同步。

对于点到点的 WAN 串行连接，两个 OSPF 路由器通常使用 HDLC 或 PPP 来形成完全邻接状态；对于以太网 LAN 连接，选取一个路由器作为 designated router（DR），再选取一个路由器作为 backup designated router（BDR），所有其他的和 DR 以及 BDR 相连的路由器形成完全邻接状态而且只传输 LSA 给 DR 和 BDR。DR 的主要功能就是保持一个 LAN 内的所有路由器拥有相同的数据库，而且把完整的数据库信息发送给新加入的路由器。路由器之间还会和 LAN 内的其他路由器（非 DR/BDR，即 DROTHERs）维持一种部分邻居关系（two-way adjacency）。

OSPF 的邻接关系一旦形成以后，会交换 LSA 来同步 LSDB，LSA 将进行可靠的洪泛。

3）OSPF 算法

OSPE 协议使用 Dijkstra 算法查找到达目标网络的最佳路径。所有的路由器拥有相同的 LSDB 后，把自己放进 SPF tree 中的 root 里，然后根据每条链路的耗费（cost），选出耗费最低的作为最佳路径，最后把最佳路径放进 forwarding database（路由表）里。图 3-2 就是一个 SPF 算法的例子。

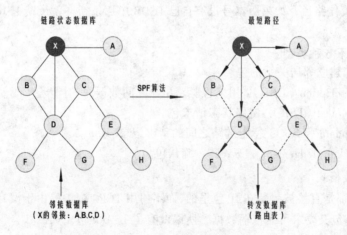

图 3-2　SPF 算法

① LSA 遵循 split horizon 原则，H 对 E 宣告它的存在，E 把 H 的宣告和它自己的宣告再传给 C 和 G；C 和 G 继续传播开来，依此类推。

② X 有 4 个邻居：A、B、C 和 D，假设这里都是以太网，每条网链路的耗费为 10，路由器可以算出最佳路径，图 3-2 的右半部分实线所标即为最佳路径。

4）链路状态结构

关于 LSA 的操作流程如图 3-3 所示，可以看出当路由器收到一个 LSA 以后，先会查看它自己的 LSDB，看有没有相应的条目，如果没有就加进自己的 LSDB 中去，并反馈 LSA 确认包（LSAck），接着再继续洪泛 LSA，最后运行 SPF 算法算出新的路由表。

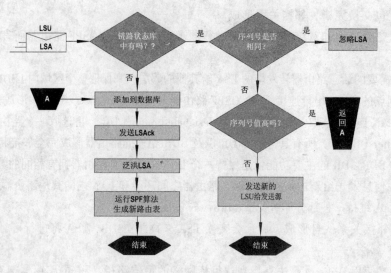

图 3-3 链路状态结构

如果在收到 LSA 的时候，自己的 LSDB 有该条目而且版本号一样，就忽略这个 LSA；如果有相应条目，但是收到的 LSA 的版本号更新，就加进自己的 LSDB 中，发回 LSAck，洪泛 LSA，最后用 SPF 计算最佳路径；如果版本号没有自己 LSDB 中那条新，就反馈 LSU 信息给发送源。

5）OSPF 包类型

OSPF 包有如下 5 种类型：
- hello：用来建立邻居关系的包。
- database description（DBD）：用来检验路由器之间数据库的同步。
- link state request（LSR）：链路状态请求包。
- link state update（LSU）：特定链路之间的请求记录。
- link state acknowledgement（LSAck）：确认包。

6）OSPF 封装格式

5 种 OSPF 包都是直接被封装在 IP 包里的，而不使用 TCP 或 UDP。由于没有使用可靠的 TCP 协议，但是 OSPF 包又要求可靠传输数据，所以就有了 LSAck 包。如图 3-4 所示就是 OSPF 包在 IP 包里的形式。

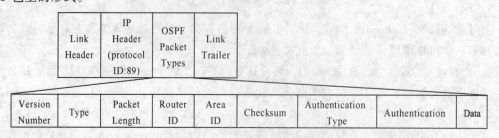

图 3-4 OSPF 封装格式

一些字段解释如下：
- Version Number：当前 OSPF 版本号。
- Type：定义 OSPF 包的类型。

- Packet Length：定义包的长度，单位字节。
- Router ID（RID）：产生 OSPF 包的源路由器。
- Area ID：定义 OSPF 包是从哪个 area 产生出来的。
- Checksum(校验和)：错误校验。
- Authentication Type：验证方法，可以是明文( cleartext )密码或者是 Message Digest 5( MD5 ) 加密格式。
- Data：对于 hello 包来说，该字段是已知邻居的列表；对于 DBD 包来说，该字段包含的是 LSDB 的汇总信息，包括 RID 等；对于 LSR 包来说，该字段包含的是需要的 LSU 类型和需要的 LSU 类型的 RID；对于 LSU 包来说，包含的是完全的 LSA 条目，多个 LSA 条目可以装在一个包里；对于 LSAck 来说，字段为空。

7）邻接（adjacency）关系建立

Hello 协议用来建立和保持 OSPF 邻接关系，采用多播地址 224.0.0.5，hello 包包含的信息如下。

① router ID（RID）：路由器的 32 位的一个唯一标识符，选取规则是，如果 loopback 接口不存在的话，就选物理接口中 IP 地址等级最高的那个，否则选取 loopback 接口。

② hello/dead intervals：定义了发送 hello 包的频率（默认在一个多路访问网络中间隔为 10 秒）；dead 间隔 4 倍于 hello 包间隔，邻居路由器之间的这些计时器必须设置成一样。

③ neighbors：邻居列表。

④ area ID：为了能够通信，OSPF 路由器的接口必须属于同一网段中的同一区域，即共享子网以及子网掩码信息。

⑤ router priority：优先级，选取 DR 和 BDR 的时候使用，8 位的一串数字。

⑥ DR/BDR IP address：DR/BDR 的 IP 地址信息。

⑦ authentication password：如果启用了验证，邻居路由器之间必须交换相同的密码信息，此项可选。

⑧ stub area flag：stub area 是通过使用默认路由代替路由更新的一种技术（功能有点像 EIGRP 中的 stub 的功能）。

8）建立双向通信

双向通信的建立过程如图 3-5 所示。刚开始 A 还没和别的路由器交换信息，还处于 down 状态，接下来通过使用多播地址 224.0.0.5 开始发送 hello 包。

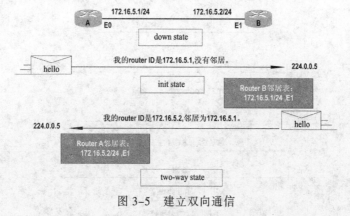

图 3-5　建立双向通信

① B接收到hello包，把A加进自己的neighbor table（邻居表）中，并进入init状态，然后以单播的形式发送hello包对A做出应答。

② A收到以后把所有从hello包里找到的RID加进自己的neighbor table中，进入two-way状态。

③ 如果链路是广播型网络，比如以太网，接下来选取DR和BDR，这一过程发生在交换信息之前。

④ 周期发送hello包保证信息交换。

9) 交换链路状态数据

选取DR和BDR后，进入exstart状态，接下来就可以发现链路状态信息并创建自己的LSDB，如图3-6所示。

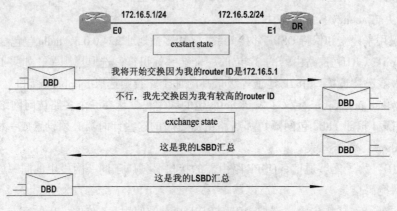

图3-6 准启动

① 在exstart状态下，邻接关系形成，路由器和DR/BDR形成主仆关系（RID等级最高的为主，其他的为辅）。

② 主辅交换DBD包（DDP），路由器进入exchange状态。

DBD包含了出现在LSDB中的LSA条目头部信息，条目信息可以为一条链路（link）或者一个网络。每个LSA条目头部信息包括链路状态类型、路由器的地址、链路耗费和序列号（版本号）。

③ 路由器收到DBD以后，将使用LSAck做出确认，还将和自己本身就有的DBD进行比较，过程如图3-7所示。

图3-7 稳定

如果 DBD 信息中有更新、更全的链路状态条目，路由器就发送 LSR 给其他路由器，该状态为 loading 状态；收到 LSR 以后，路由器做出响应，以 LSU 作为应答，其中包含了 LSR 所需要的完整信息；收到 LSU 以后，再次做出确认，发送 LSAck。

④ 路由器添加新的条目到 LSDB 中，进入 full 状态，接下来就可以对数据进行路由了。

10）路由更新

在链路状态发生变化以后，路由器将洪泛 LSA 来对其他路由器做出通知，如图 3-8 所示。

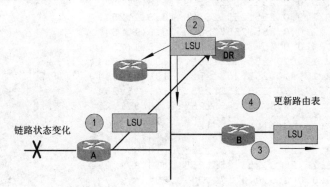

图 3-8　路由更新

① 路由器意识到链路产生变化以后，对多播地址 224.0.0.6 和所有的 DR/BDR 发送 LSU，其中 LSU 包含了更新了的 LSA 条目。

② DR 对 LSU 做出确认，接着对多播地址 224.0.0.5 继续洪泛，每个收到 LSU 的路由器对 DR 做出确认（反馈 LSAck）。

③ 如果路由器连接了其他网络，将通过转发 LSU 给 DR（在点到点网络是转发给邻居路由器）来对其他网络进行洪泛。

④ 其他路由器通过 LSU 来更新自己的 LSDB，然后使用 SPF 算法重新计算最佳路径。

⑤ 链路状态条目的最大生存周期是 60 分钟，60 分钟以后，它将从 LSDB 中被移除。

## 项目实施

1）使用静态路由配置全网

① 在汇聚交换机、核心交换机、路由器上配置静态路由联通全网。

② 创建核心层设备配置脚本。

CORE1 脚本为 CORE1_STATIC R_CFG.txt，内容如下：

```
################### core1-static r-cfg ###################
<省略 VLAN 部分>
ip route 172.20.1.0 255.255.255.0 10.8.0.254
ip route 10.8.3.0 255.255.255.0 10.8.0.254
ip route 172.16.0.0 255.255.255.0 10.8.2.254
ip route 10.10.0.0 255.255.255.0 10.8.2.254
ip route 10.11.0.0 255.255.255.0 10.8.2.254
ip route 10.12.0.0 255.255.255.0 10.8.2.254
ip route 10.13.0.0 255.255.255.0 10.8.2.254
ip route 172.16.0.0 255.255.255.0 10.8.2.254
```

```
ip route 172.16.4.0 255.255.255.0 10.8.2.254
ip route 172.16.8.0 255.255.255.0 10.8.2.254
ip route 172.16.12.0 255.255.255.0 10.8.2.254
ip route 172.16.16.0 255.255.255.0 10.8.2.254
ip route 172.16.20.0 255.255.255.0 10.8.2.254
ip route 172.16.24.0 255.255.255.0 10.8.2.254
ip route 172.16.28.0 255.255.255.0 10.8.2.254
ip route 172.16.32.0 255.255.255.0 10.8.2.254
ip route 10.20.0.0 255.255.255.0 10.8.1.254
ip route 10.21.0.0 255.255.255.0 10.8.1.254
ip route 10.22.0.0 255.255.255.0 10.8.1.254
ip route 10.23.0.0 255.255.255.0 10.8.1.254
ip route 172.17.0.0 255.255.255.0 10.8.1.254
ip route 172.17.4.0 255.255.255.0 10.8.1.254
ip route 172.17.8.0 255.255.255.0 10.8.1.254
ip route 172.17.12.0 255.255.255.0 10.8.1.254
ip route 172.17.16.0 255.255.255.0 10.8.1.254
ip route 172.17.20.0 255.255.255.0 10.8.1.254
ip route 172.17.24.0 255.255.255.0 10.8.1.254
ip route 172.17.28.0 255.255.255.0 10.8.1.254
ip route 0.0.0.0 0.0.0.0 172.30.0.254
ip route 172.20.0.0 255.255.0.0 10.8.0.254
exit
```

CORE2 脚本为 CORE2_STATIC R_CFG.txt，内容如下：

```
##################     core2-static r-cfg     ##################
<省略 VLAN 部分>
ip route 172.20.1.0 255.255.255.0 10.8.1.1
ip route 172.20.2.0 255.255.255.0 10.8.1.1
ip route 10.8.2.0 255.255.255.0 10.8.1.1
ip route 10.10.0.0 255.255.255.0 10.8.3.254
ip route 10.11.0.0 255.255.255.0 10.8.3.254
ip route 10.12.0.0 255.255.255.0 10.8.3.254
ip route 10.13.0.0 255.255.255.0 10.8.3.254
ip route 172.16.0.0 255.255.255.0 10.8.3.254
ip route 172.16.4.0 255.255.255.0 10.8.3.254
ip route 172.16.8.0 255.255.255.0 10.8.3.254
ip route 172.16.12.0 255.255.255.0 10.8.3.254
ip route 172.16.16.0 255.255.255.0 10.8.3.254
ip route 172.16.20.0 255.255.255.0 10.8.3.254
ip route 172.16.24.0 255.255.255.0 10.8.3.254
ip route 172.16.28.0 255.255.255.0 10.8.3.254
ip route 172.16.32.0 255.255.255.0 10.8.3.254
ip route 172.17.0.0 255.255.255.0 10.20.0.254
ip route 172.17.4.0 255.255.255.0 10.20.0.254
ip route 172.17.8.0 255.255.255.0 10.21.0.254
ip route 172.17.12.0 255.255.255.0 10.21.0.254
ip route 172.17.16.0 255.255.255.0 10.22.0.254
ip route 172.17.20.0 255.255.255.0 10.22.0.254
ip route 172.17.24.0 255.255.255.0 10.23.0.254
ip route 172.17.28.0 255.255.255.0 10.23.0.254
ip route 172.20.0.0 255.255.0.0 10.8.1.1
ip route 0.0.0.0 0.0.0.0 172.30.1.254
exit
```

CORE3 脚本为 CORE3_STATIC R_CFG.txt，内容如下：

```
##################     core3-static r-cfg     ##################
<省略 VLAN 部分>
ip route 172.20.0.0 255.255.0.0 10.8.2.1
ip route 10.8.1.0 255.255.255.0 10.8.2.1
ip route 10.20.0.0 255.255.255.0 10.8.3.1
ip route 10.21.0.0 255.255.255.0 10.8.3.1
ip route 10.22.0.0 255.255.255.0 10.8.3.1
ip route 10.23.0.0 255.255.255.0 10.8.3.1
ip route 172.17.0.0 255.255.0.0 10.8.3.1
ip route 172.16.0.0 255.255.255.0 10.10.0.254
ip route 172.16.4.0 255.255.255.0 10.10.0.254
ip route 172.16.8.0 255.255.255.0 10.11.0.254
ip route 172.16.12.0 255.255.255.0 10.11.0.254
ip route 172.16.16.0 255.255.255.0 10.12.0.254
ip route 172.16.20.0 255.255.255.0 10.12.0.254
ip route 172.16.24.0 255.255.255.0 10.13.0.254
ip route 172.16.28.0 255.255.255.0 10.13.0.254
ip route 172.16.32.0 255.255.255.0 10.13.0.254
ip route 0.0.0.0 0.0.0.0 10.8.2.1
exit
```

CORE4 脚本为 CORE4_STATIC R_CFG.txt，内容如下：

```
##################     core4-static r-cfg     ##################
<省略 VLAN 部分>
ip route 0.0.0.0 0.0.0.0 10.8.0.1
exit
```

③ 创建汇聚层设备配置脚本。

DS_1_GY 脚本为 DS_1_GY_STATIC R_CFG.txt，内容如下：

```
##################     ds-1-gy-static r-cfg     ##################
<省略 VLAN 部分>
ip route 0.0.0.0 0.0.0.0 10.10.0.1
exit
```

DS_2_GY 脚本为 DS_2_GY_STATIC R_CFG.txt，内容如下：

```
##################     ds-2-gy-static r-cfg     ##################
<省略 VLAN 部分>
ip route 0.0.0.0 0.0.0.0 11.10.0.1
exit
```

DS_3_GY 脚本为 DS_3_GY_STATIC R_CFG.txt，内容如下：

```
##################     ds-3-gy-static r-cfg     ##################
<省略 VLAN 部分>
ip route 0.0.0.0 0.0.0.0 12.10.0.1
exit
```

DS_4_GY 脚本为 DS_4_GY_STATIC R_CFG.txt，内容如下：

```
##################     ds-4-gy-static r-cfg     ##################
<省略 VLAN 部分>
ip route 0.0.0.0 0.0.0.0 13.10.0.1
exit
```

DS_JX 脚本为 DS_JX_STATIC R_CFG.txt，内容如下：

```
##################    ds-jx-static r-cfg    ##################
<省略 VLAN 部分>
ip route 0.0.0.0 0.0.0.0 10.20.0.1
exit
```

DS_SX 脚本为 DS_SX_STATIC R_CFG.txt，内容如下：

```
##################    ds-sx-static r-cfg    ##################
<省略 VLAN 部分>
ip route 0.0.0.0 0.0.0.0 10.21.0.1
exit
```

DS_TSG 脚本为 DS_TSG_STATIC R_CFG.txt，内容如下：

```
##################    ds-tsg-static r-cfg    ##################
<省略 VLAN 部分>
ip route 0.0.0.0 0.0.0.0 10.22.0.1
exit
```

DS_ZH 脚本为 DS_ZH_STATIC R_CFG.txt，内容如下：

```
##################    ds-zh-static r-cfg    ##################
<省略 VLAN 部分>
ip route 0.0.0.0 0.0.0.0 10.23.0.1
exit
```

④ 配置 BR。

BR 脚本为 BR_STATIC R_CFG.txt，内容如下：

```
##################    br-static r-cfg    ##################
<省略 VLAN 部分>
ip route 172.16.0.0 255.255.0.0 172.30.0.1
ip route 172.17.0.0 255.255.0.0 172.30.0.1
ip route 172.17.0.0 255.255.0.0 172.31.0.1
ip route 172.16.0.0 255.255.0.0 172.31.0.1
ip route 172.20.0.0 255.255.0.0 172.30.0.1
ip route 10.0.0.0 255.0.0.0 172.30.0.1
ip route 0.0.0.0 0.0.0.0 58.0.0.1
exit
```

⑤ 在各个设备上执行脚本。

⑥ 在 CORE1 上使用 show ip route 命令查看路由表：

```
core1#show ip route
Codes: C - connected, S - static, I - IGRP, R - RIP, M - mobile, B - BGP
       D - EIGRP, EX - EIGRP external, O - OSPF, IA - OSPF inter area
       N1 - OSPF NSSA external type 1, N2 - OSPF NSSA external type 2
       E1 - OSPF external type 1, E2 - OSPF external type 2, E - EGP
       i - IS-IS, L1 - IS-IS level-1, L2 - IS-IS level-2, ia - IS-IS inter
       area
       * - candidate default, U - per-user static route, o - ODR
       P - periodic downloaded static route
Gateway of last resort is 172.30.0.254 to network 0.0.0.0


     10.0.0.0/24 is subnetted, 12 subnets
```

```
C       10.8.0.0 is directly connected, Vlan300
C       10.8.1.0 is directly connected, Vlan301
C       10.8.2.0 is directly connected, Vlan302
S       10.8.3.0 [1/0] via 10.8.0.254
S       10.10.0.0 [1/0] via 10.8.2.254
S       10.11.0.0 [1/0] via 10.8.2.254
S       10.12.0.0 [1/0] via 10.8.2.254
S       10.13.0.0 [1/0] via 10.8.2.254
S       10.20.0.0 [1/0] via 10.8.1.254
S       10.21.0.0 [1/0] via 10.8.1.254
S       10.22.0.0 [1/0] via 10.8.1.254
S       10.23.0.0 [1/0] via 10.8.1.254
        172.16.0.0/24 is subnetted, 9 subnets
S       172.16.0.0 [1/0] via 10.8.2.254
S       172.16.4.0 [1/0] via 10.8.2.254
S       172.16.8.0 [1/0] via 10.8.2.254
S       172.16.12.0 [1/0] via 10.8.2.254
S       172.16.16.0 [1/0] via 10.8.2.254
S       172.16.20.0 [1/0] via 10.8.2.254
S       172.16.24.0 [1/0] via 10.8.2.254
S       172.16.28.0 [1/0] via 10.8.2.254
S       172.16.32.0 [1/0] via 10.8.2.254
        172.17.0.0/24 is subnetted, 8 subnets
S       172.17.0.0 [1/0] via 10.8.1.254
S       172.17.4.0 [1/0] via 10.8.1.254
S       172.17.8.0 [1/0] via 10.8.1.254
S       172.17.12.0 [1/0] via 10.8.1.254
S       172.17.16.0 [1/0] via 10.8.1.254
S       172.17.20.0 [1/0] via 10.8.1.254
S       172.17.24.0 [1/0] via 10.8.1.254
S       172.17.28.0 [1/0] via 10.8.1.254
        172.20.0.0/16 is variably subnetted, 2 subnets, 2 masks
S       172.20.0.0/16 [1/0] via 10.8.0.254
S       172.20.1.0/24 [1/0] via 10.8.0.254
        172.30.0.0/24 is subnetted, 1 subnets
C       172.30.0.0 is directly connected, FastEthernet0/24
S*      0.0.0.0/0 [1/0] via 172.30.0.254
```

⑦ 在 CORE2 上使用 show ip route 命令查看路由表：

```
Core2#show ip route
Codes: C - connected, S - static, I - IGRP, R - RIP, M - mobile, B - BGP
       D - EIGRP, EX - EIGRP external, O - OSPF, IA - OSPF inter area
       N1 - OSPF NSSA external type 1, N2 - OSPF NSSA external type 2
       E1 - OSPF external type 1, E2 - OSPF external type 2, E - EGP
       i - IS-IS, L1 - IS-IS level-1, L2 - IS-IS level-2, ia - IS-IS inter
       area
       * - candidate default, U - per-user static route, o - ODR
```

```
         P - periodic downloaded static route

Gateway of last resort is 172.30.1.254 to network 0.0.0.0

     10.0.0.0/24 is subnetted, 11 subnets
C       10.8.1.0 is directly connected, Vlan301
S       10.8.2.0 [1/0] via 10.8.1.1
C       10.8.3.0 is directly connected, Vlan303
S       10.10.0.0 [1/0] via 10.8.3.254
S       10.11.0.0 [1/0] via 10.8.3.254
S       10.12.0.0 [1/0] via 10.8.3.254
S       10.13.0.0 [1/0] via 10.8.3.254
C       10.20.0.0 is directly connected, Vlan320
C       10.21.0.0 is directly connected, Vlan321
C       10.22.0.0 is directly connected, Vlan322
C       10.23.0.0 is directly connected, Vlan323
     172.16.0.0/24 is subnetted, 9 subnets
S       172.16.0.0 [1/0] via 10.8.3.254
S       172.16.4.0 [1/0] via 10.8.3.254
S       172.16.8.0 [1/0] via 10.8.3.254
S       172.16.12.0 [1/0] via 10.8.3.254
S       172.16.16.0 [1/0] via 10.8.3.254
S       172.16.20.0 [1/0] via 10.8.3.254
S       172.16.24.0 [1/0] via 10.8.3.254
S       172.16.28.0 [1/0] via 10.8.3.254
S       172.16.32.0 [1/0] via 10.8.3.254
     172.17.0.0/24 is subnetted, 8 subnets
S       172.17.0.0 [1/0] via 10.20.0.254
S       172.17.4.0 [1/0] via 10.20.0.254
S       172.17.8.0 [1/0] via 10.21.0.254
S       172.17.12.0 [1/0] via 10.21.0.254
S       172.17.16.0 [1/0] via 10.22.0.254
S       172.17.20.0 [1/0] via 10.22.0.254
S       172.17.24.0 [1/0] via 10.23.0.254
S       172.17.28.0 [1/0] via 10.23.0.254
     172.20.0.0/16 is variably subnetted, 3 subnets, 2 masks
S       172.20.0.0/16 [1/0] via 10.8.1.1
S       172.20.1.0/24 [1/0] via 10.8.1.1
S       172.20.2.0/24 [1/0] via 10.8.1.1
     172.30.0.0/24 is subnetted, 1 subnets
C       172.30.1.0 is directly connected, FastEthernet0/24
S*   0.0.0.0/0 [1/0] via 172.30.1.254
```

⑧ 在 CORE3 上使用 show ip route 命令查看路由表：

```
core3#show ip route
Codes: C - connected, S - static, I - IGRP, R - RIP, M - mobile, B - BGP
       D - EIGRP, EX - EIGRP external, O - OSPF, IA - OSPF inter area
```

```
              N1 - OSPF NSSA external type 1, N2 - OSPF NSSA external type 2
              E1 - OSPF external type 1, E2 - OSPF external type 2, E - EGP
              i - IS-IS, L1 - IS-IS level-1, L2 - IS-IS level-2, ia - IS-IS inter
              area
              * - candidate default, U - per-user static route, o - ODR
              P - periodic downloaded static route

Gateway of last resort is 10.8.2.1 to network 0.0.0.0

        10.0.0.0/24 is subnetted, 11 subnets
S       10.8.1.0 [1/0] via 10.8.2.1
C       10.8.2.0 is directly connected, Vlan302
C       10.8.3.0 is directly connected, Vlan303
C       10.10.0.0 is directly connected, Vlan310
C       10.11.0.0 is directly connected, Vlan311
C       10.12.0.0 is directly connected, Vlan312
C       10.13.0.0 is directly connected, Vlan313
S       10.20.0.0 [1/0] via 10.8.3.1
S       10.21.0.0 [1/0] via 10.8.3.1
S       10.22.0.0 [1/0] via 10.8.3.1
S       10.23.0.0 [1/0] via 10.8.3.1
        172.16.0.0/24 is subnetted, 9 subnets
S       172.16.0.0 [1/0] via 10.10.0.254
S       172.16.4.0 [1/0] via 10.10.0.254
S       172.16.8.0 [1/0] via 10.11.0.254
S       172.16.12.0 [1/0] via 10.11.0.254
S       172.16.16.0 [1/0] via 10.12.0.254
S       172.16.20.0 [1/0] via 10.12.0.254
S       172.16.24.0 [1/0] via 10.13.0.254
S       172.16.28.0 [1/0] via 10.13.0.254
S       172.16.32.0 [1/0] via 10.13.0.254
S    172.17.0.0/16 [1/0] via 10.8.3.1
S    172.20.0.0/16 [1/0] via 10.8.2.1
S*   0.0.0.0/0 [1/0] via 10.8.2.1
```

⑨ 分析：可以发现静态路由在配置过程中工作量比较大，而且容易出错，因为网络中网段（子网）太多，使用静态路由联通校园网显然不是明智的选择。

2）项目设计

图3-9所示为校园网路由简化图。

在内网运行OSPF，均在area0中，分析LSA的传播范围带来的问题。

为了限制LSA传播范围，重新设计区域，核心网络在area 0中，宿舍楼处于area 10，教学楼等处于area 20。

为控制LSA的不同类型传播，优化网络路由流量，在网络中设计stub、totally stub、nssa等，最终确定校园网OSPF设计最佳方案。

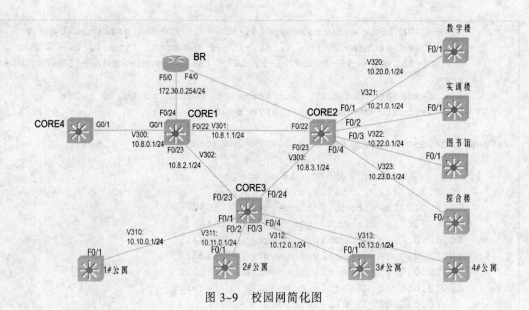

图 3-9　校园网简化图

3）项目实施

使用 OSPF 部署校园网路由，要求产生最优方案，同时发布默认路由。

（1）单区域（area 0）实现

① 校园网全网（除边界 BR 外）运行 OSPF，均处于 area 0 中。

② 核心层配置命令如下。

配置 CORE1：

```
CORE_1 (config)#ip route 0.0.0.0 0.0.0.0 172.30.1.254
CORE_1 (config)#router ospf 100
CORE_1 (config-router)#router-id 10.5.0.2
CORE_1 (config-router)#redistribute connected subnets
CORE_1 (config-router)#network 10.8.1.0 0.0.0.255 area 0
CORE_1 (config-router)#network 10.8.3.0 0.0.0.255 area 0
CORE_1 (config-router)#network 10.20.0.0 0.0.0.255 area 0
CORE_1 (config-router)#network 10.21.0.0 0.0.0.255 area 0
CORE_1 (config-router)#network 10.22.0.0 0.0.0.255 area 0
CORE_1 (config-router)#network 10.23.0.0 0.0.0.255 area 0
CORE_1 (config-router)#default-information originate
CORE_1 (config-router)#exit
```

配置 CORE2：

```
CORE_2 (config)#ip route 0.0.0.0 0.0.0.0 172.30.1.254
CORE_2 (config)#router ospf 100
CORE_2 (config-router)#router-id 10.5.0.2
CORE_2 (config-router)#redistribute connected subnets
CORE_2 (config-router)#network 10.8.1.0 0.0.0.255 area 0
CORE_2 (config-router)#network 10.8.3.0 0.0.0.255 area 0
CORE_2 (config-router)#network 10.20.0.0 0.0.0.255 area 0
CORE_2 (config-router)#network 10.21.0.0 0.0.0.255 area 0
CORE_2 (config-router)#network 10.22.0.0 0.0.0.255 area 0
CORE_2 (config-router)#network 10.23.0.0 0.0.0.255 area 0
```

```
CORE_2 (config-router)#default-information originate
CORE_2 (config-router)#exit
```

配置 CORE3：

```
CORE_3 (config)# router ospf 100
CORE_3 (config-router)# router-id 10.5.0.3
CORE_3 (config-router)# redistribute connected subnets
CORE_3 (config-router)# network 10.8.2.0 0.0.0.255 area 0
CORE_3 (config-router)# network 10.8.3.0 0.0.0.255 area 0
CORE_3 (config-router)# network 10.10.0.0 0.0.0.255 area 0
CORE_3 (config-router)# network 10.11.0.0 0.0.0.255 area 0
CORE_3 (config-router)# network 10.12.0.0 0.0.0.255 area 0
CORE_3 (config-router)# network 10.13.0.0 0.0.0.255 area 0
CORE_3 (config-router)# exit
```

配置 CORE4：

```
CORE_4(config)# router ospf 100
CORE_4(config-router)# router-id 10.5.0.4
CORE_4(config-router)#redistribute connected subnets
CORE_4(config-router)#network 172.20.0.0 0.0.255.255 area 0
CORE_4(config-router)#network 10.8.0.0 0.0.0.255 area 0
CORE_4(config-router)#exit
```

③ 汇聚层配置命令如下。

配置 DS_1_GY：

```
DS_1_GY (config)# router ospf 100
DS_1_GY (config-router)# router-id 10.10.0.1
DS_1_GY (config-router)#redistribute connected subnets
DS_1_GY (config-router)#network 10.10.0.0 0.0.0.255 area 0
DS_1_GY (config-router)#network 172.16.0.0 0.0.0.255 area 0
DS_1_GY (config-router)#network 172.16.4.0 0.0.0.255 area 0
DS_1_GY (config-router)#exit
```

配置 DS_2_GY：

```
DS_2_GY (config)# router ospf 100
DS_2_GY (config-router)#router-id 10.10.0.2
DS_2_GY (config-router)#redistribute connected subnets
DS_2_GY (config-router)#network 10.11.0.0 0.0.0.255 area 0
DS_2_GY (config-router)#network 172.16.8.0 0.0.0.255 area 0
DS_2_GY (config-router)#network 172.16.12.0 0.0.0.255 area 0
DS_2_GY (config-router)#exit
```

配置 DS_3_GY：

```
DS_3_GY (config)# router ospf 100
DS_3_GY (config-router)# router-id 10.10.0.3
DS_3_GY (config-router)#redistribute connected subnets
DS_3_GY (config-router)#network 10.12.0.0 0.0.0.255 area 0
DS_3_GY (config-router)#network 172.16.16.0 0.0.0.255 area 0
DS_3_GY (config-router)#network 172.16.20.0 0.0.0.255 area 0
DS_3_GY (config-router)#exit
```

配置 DS_4_GY：

```
DS_4_GY (config)# router ospf 100
DS_4_GY (config-router)#router-id 10.10.0.4
DS_4_GY (config-router)#redistribute connected subnets
DS_4_GY (config-router)#network 10.13.0.0 0.0.0.255 area 0
DS_4_GY (config-router)#network 172.16.24.0 0.0.0.255 area 0
DS_4_GY (config-router)#network 172.16.28.0 0.0.0.255 area 0
DS_4_GY (config-router)#network 172.16.32.0 0.0.0.255 area 0
DS_4_GY (config-router)#exit
```

配置 DS_JX：

```
DS_JX (config)# router ospf 100
DS_JX (config-router)# router-id 10.20.0.1
DS_JX (config-router)# redistribute connected subnets
DS_JX (config-router)# network 10.20.0.0 0.0.0.255 area 0
DS_JX (config-router)# network 172.17.0.0 0.0.0.255 area 0
DS_JX (config-router)# network 172.17.4.0 0.0.0.255 area 0
DS_JX (config-router)# exit
```

配置 DS_SX：

```
DS_SX(config)# router ospf 100
DS_SX(config-router)# router-id 10.20.0.2
DS_SX(config-router)# redistribute connected subnets
DS_SX(config-router)# network 10.21.0.0 0.0.0.255 area 0
DS_SX(config-router)# network 172.17.8.0 0.0.0.255 area 0
DS_SX(config-router)# network 172.17.12.0 0.0.0.255 area 0
DS_SX(config-router)# exit
```

配置 DS_TSG：

```
DS_TSG(config)# router ospf 100
DS_TSG(config-router)# router ospf 100
DS_TSG(config-router)# router-id 10.20.0.3
DS_TSG(config-router)# redistribute connected subnets
DS_TSG(config-router)# network 10.22.0.0 0.0.0.255 area 0
DS_TSG(config-router)# network 172.17.16.0 0.0.0.255 area 0
DS_TSG(config-router)# network 172.17.20.0 0.0.0.255 area 0
DS_TSG(config-router)# exit
```

配置 DS_ZH：

```
DS_ZH(config)# router ospf 100
DS_ZH(config-router)#router-id 10.20.0.4
DS_ZH(config-router)#redistribute connected subnets
DS_ZH(config-router)#network 10.23.0.0 0.0.0.255 area 0
DS_ZH(config-router)#network 172.17.24.0 0.0.0.255 area 0
DS_ZH(config-router)#network 172.17.28.0 0.0.0.255 area 0
DS_ZH(config-router)#network 172.17.32.0 0.0.0.255 area 0
DS_ZH(config-router)#exit
```

④ 在汇聚层交换机 DS_1_GY 上使用 show ip route 查看路由表：

```
DS_1_GY#sh ip route
    10.0.0.0/24 is subnetted, 12 subnets
O      10.8.0.0 [110/3] via 10.10.0.1, 00:03:32, Vlan310
O      10.8.1.0 [110/3] via 10.10.0.1, 00:03:32, Vlan310
```

```
O       10.8.2.0 [110/2] via 10.10.0.1, 00:03:32, Vlan310
O       10.8.3.0 [110/2] via 10.10.0.1, 00:03:32, Vlan310
C       10.10.0.0 is directly connected, Vlan310
O       10.11.0.0 [110/2] via 10.10.0.1, 00:03:32, Vlan310
O       10.12.0.0 [110/2] via 10.10.0.1, 00:03:32, Vlan310
O       10.13.0.0 [110/2] via 10.10.0.1, 00:03:32, Vlan310
O       10.20.0.0 [110/3] via 10.10.0.1, 00:03:32, Vlan310
O       10.21.0.0 [110/3] via 10.10.0.1, 00:03:32, Vlan310
O       10.22.0.0 [110/3] via 10.10.0.1, 00:03:32, Vlan310
O       10.23.0.0 [110/3] via 10.10.0.1, 00:03:32, Vlan310
     172.16.0.0/24 is subnetted, 9 subnets
C       172.16.0.0 is directly connected, FastEthernet0/23
C       172.16.4.0 is directly connected, Vlan110
O       172.16.8.0 [110/3] via 10.10.0.1, 00:02:38, Vlan310
O       172.16.12.0 [110/3] via 10.10.0.1, 00:02:38, Vlan310
O       172.16.16.0 [110/3] via 10.10.0.1, 00:02:38, Vlan310
O       172.16.20.0 [110/3] via 10.10.0.1, 00:02:38, Vlan310
O       172.16.24.0 [110/3] via 10.10.0.1, 00:02:13, Vlan310
O       172.16.28.0 [110/3] via 10.10.0.1, 00:02:13, Vlan310
O       172.16.32.0 [110/20] via 10.10.0.1, 00:02:13, Vlan310
     172.17.0.0/24 is subnetted, 8 subnets
O       172.17.0.0 [110/4] via 10.10.0.1, 00:01:57, Vlan310
O       172.17.4.0 [110/4] via 10.10.0.1, 00:01:57, Vlan310
O       172.17.8.0 [110/4] via 10.10.0.1, 00:00:42, Vlan310
O       172.17.12.0 [110/4] via 10.10.0.1, 00:00:42, Vlan310
O       172.17.16.0 [110/4] via 10.10.0.1, 00:00:32, Vlan310
O       172.17.20.0 [110/4] via 10.10.0.1, 00:00:32, Vlan310
O       172.17.24.0 [110/4] via 10.10.0.1, 00:00:32, Vlan310
O       172.17.28.0 [110/4] via 10.10.0.1, 00:00:32, Vlan310
     172.20.0.0/24 is subnetted, 2 subnets
O       172.20.1.0 [110/4] via 10.10.0.1, 00:03:32, Vlan310
O       172.20.2.0 [110/4] via 10.10.0.1, 00:03:32, Vlan310
     172.30.0.0/24 is subnetted, 2 subnets
O E2    172.30.0.0 [110/20] via 10.10.0.1, 00:03:32, Vlan310
O E2    172.30.1.0 [110/20] via 10.10.0.1, 00:03:32, Vlan310
O*E2 0.0.0.0/0 [110/1] via 10.10.0.1, 00:03:32, Vlan310
```

⑤ 仔细检查，汇聚层交换机已经学习到了全网路由：

O E2 类型的路由为 OSPF 使用 "redistribute connected subnets" 发布的直连路由。

O*E2 类型的路由为学习到核心层 "default-information originate" 发布的直连路由。

⑥ 在汇聚层交换机 DS_1_GY 上使用 show ip ospf database 查看链路状态数据库：

```
DS_1_GY#show ip ospf database
            Router Link States (Area 0)    !!! LSA 类型 1 用于宣告具体链路
Link ID         ADV Router      Age         Seq#       Checksum Link count
10.5.0.4        10.5.0.4        1642        0x80000006 0x003728 3
10.5.0.1        10.5.0.1        1628        0x80000007 0x008d0b 3
10.10.0.1       10.10.0.1       1375        0x80000006 0x0072eb 3
10.10.0.2       10.10.0.2       1336        0x80000006 0x002822 3
10.10.0.3       10.10.0.3       1326        0x80000006 0x00dd58 3
10.5.0.3        10.5.0.3        1306        0x8000000d 0x00c50a 6
10.20.0.1       10.20.0.1       1290        0x80000006 0x00a68d 3
10.20.0.2       10.20.0.2       1221        0x80000006 0x005cc3 3
```

```
10.20.0.3        10.20.0.3        1210     0x80000006 0x0012f9  3
10.5.0.2         10.5.0.2         1210     0x8000000d 0x00443d  6
10.20.0.4        10.20.0.4        1210     0x80000005 0x00c92f  3
10.10.0.4        10.10.0.4        383      0x80000006 0x00f041  4
            Net Link States (Area 0)              !!! LSA 类型 2 用于宣告具体网络
Link ID          ADV Router       Age      Seq#       Checksum
10.8.0.1         10.5.0.1         1642     0x80000001 0x00309d
10.8.1.254       10.5.0.2         1642     0x80000001 0x00e730
10.8.3.254       10.5.0.3         1628     0x80000001 0x00f7dd
10.8.2.254       10.5.0.3         1628     0x80000002 0x009cfc
10.10.0.1        10.5.0.3         1375     0x80000003 0x008065
10.11.0.1        10.5.0.3         1335     0x80000004 0x0013c1
10.12.0.1        10.5.0.3         1326     0x80000005 0x00ca3b
10.13.0.1        10.5.0.3         1306     0x80000006 0x00c240
10.20.0.1        10.5.0.2         1290     0x80000002 0x000bcc
10.21.0.1        10.5.0.2         1220     0x80000003 0x008a45
10.22.0.1        10.5.0.2         1211     0x80000004 0x0004a8
10.23.0.1        10.5.0.2         1211     0x80000005 0x007081
            Type-5 AS External Link States        !!! LSA 类型 5
Link ID          ADV Router       Age      Seq#       Checksum  Tag
172.30.1.0       10.5.0.2         1681     0x80000006 0x00f7e6  0
0.0.0.0          10.5.0.2         1681     0x80000007 0x008239  1
172.30.0.0       10.5.0.1         1696     0x80000003 0x000fd4  0
0.0.0.0          10.5.0.1         1696     0x80000004 0x008e31  1
172.16.32.0      10.10.0.4        381      0x80000005 0x0018af  0
```

⑦ 在 DS_JX 交换机将 F0/1 口人为 shutdown，模拟故障。

⑧ 在仿真模式下，过滤数据包，仅分析 OSPF，在 CORE2 查看 LSA 更新包，注意目标地址为 255.0.0.5，详见图 3-10。

⑨ 分析 LSU 封装内容，详见图 3-11。

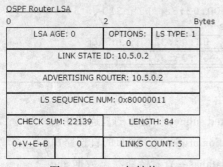

图 3-10  LSA 包结构

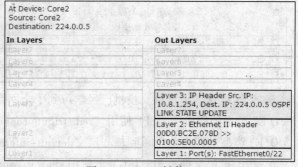

图 3-11  LSU 封装

⑩ LSA 包通过组播方式发送给其他所有路由器。当 LSA 更新包到达 DS_1_GY 时，路由表中将删除 DS_JX 所通告的路由。

总结：OSPF 路由是通过 LSA 来进行通告的。在单区域部署 OSPF 时，会导致 LSA 在全网传递，从而降低网络的利用率。

（2）区域设计

① 设计 area 0，核心设备在 area 0 运行 OSPF。

② 为有效控制 LSA 的传播范围，其他区域，如宿舍楼运行在 area 10，教学楼、实训楼、

图书馆、综合楼运行在 area20，见图 3-12。

③ 核心层配置如下。

配置 CORE1：

```
CORE_1 (config)# ip route 0.0.0.0 0.0.0.0 172.30.0.254
CORE_1 (config-router)#router ospf 100
CORE_1 (config-router)#router-id 10.5.0.1
CORE_1 (config-router)#redistribute connected subnets
CORE_1 (config-router)#network 10.8.0.0 0.0.0.255 area 0
CORE_1 (config-router)#network 10.8.1.0 0.0.0.255 area 0
CORE_1 (config-router)#network 10.8.2.0 0.0.0.255 area 0
CORE_1 (config-router)#default-information originate
CORE_1 (config-router)#exit
```

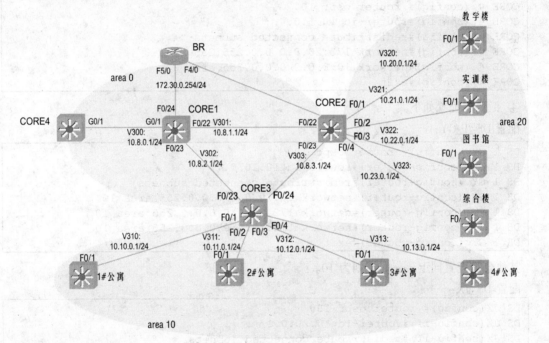

图 3-12　区域设计图

配置 CORE2：

```
CORE_2 (config)# ip route 0.0.0.0 0.0.0.0 172.30.1.254
CORE_2 (config-router)# router ospf 100
CORE_2 (config-router)#router-id 10.5.0.2
CORE_2 (config-router)#redistribute connected subnets
CORE_2 (config-router)#network 10.8.1.0 0.0.0.255 area 0
CORE_2 (config-router)#network 10.8.3.0 0.0.0.255 area 0
CORE_2 (config-router)#network 10.20.0.0 0.0.0.255 area 20
CORE_2 (config-router)#network 10.21.0.0 0.0.0.255 area 20
CORE_2 (config-router)#network 10.22.0.0 0.0.0.255 area 20
CORE_2 (config-router)#network 10.23.0.0 0.0.0.255 area 20
CORE_2 (config-router)#default-information originate
CORE_2 (config-router)#exit
```

配置 CORE3：

```
CORE_3 (config)# router ospf 100
CORE_3 (config-router)# router-id 10.5.0.3
CORE_3 (config-router)#redistribute connected subnets
CORE_3 (config-router)#network 10.8.2.0 0.0.0.255 area 0
CORE_3 (config-router)#network 10.8.3.0 0.0.0.255 area 0
CORE_3 (config-router)#network 10.10.0.0 0.0.0.255 area 10
CORE_3 (config-router)#network 10.11.0.0 0.0.0.255 area 10
CORE_3 (config-router)#network 10.12.0.0 0.0.0.255 area 10
CORE_3 (config-router)#network 10.13.0.0 0.0.0.255 area 10
exit
```

配置 CORE4：

```
CORE_4 (config)# router ospf 100
CORE_4 (config)#router-id 10.5.0.4
CORE_4 (config)#redistribute connected subnets
CORE_4 (config)#network 172.20.0.0 0.0.255.255 area 0
CORE_4 (config)#network 10.8.0.0 0.0.0.255 area 0
CORE_4 (config)#exit
```

④ 汇聚层配置如下。

配置 DS_1_GY：

```
DS_1_GY (config)# router ospf 100
DS_1_GY (config-router)#router-id 10.10.0.1
DS_1_GY (config-router)#redistribute connected subnets
DS_1_GY (config-router)#network 10.10.0.0 0.0.0.255 area 10
DS_1_GY (config-router)#network 172.16.0.0 0.0.0.255 area 10
DS_1_GY (config-router)#network 172.16.4.0 0.0.0.255 area 10
DS_1_GY (config-router)#exit
```

其他公寓配置同上，area 规划为 10。

配置 DS_JX：

```
DS_JX(config)# router ospf 100
DS_JX(config-if)#router-id 10.20.0.1
DS_JX(config-if)#redistribute connected subnets
DS_JX(config-if)#network 10.20.0.0 0.0.0.255 area 20
DS_JX(config-if)#network 172.17.0.0 0.0.0.255 area 20
DS_JX(config-if)#network 172.17.4.0 0.0.0.255 area 20
DS_JX(config-if)#exit
```

其他楼宇配置同上，area 规划为 20。

⑤ 在 CORE1 上使用 show ip route 命令查看路由表。

```
core1#sh ip route
     10.0.0.0/24 is subnetted, 12 subnets
C       10.8.0.0 is directly connected, Vlan300
C       10.8.1.0 is directly connected, Vlan301
C       10.8.2.0 is directly connected, Vlan302
O       10.8.3.0 [110/2] via 10.8.1.254, 00:02:38, Vlan301
                 [110/2] via 10.8.2.254, 00:02:28, Vlan302
O IA    10.10.0.0 [110/2] via 10.8.2.254, 00:01:22, Vlan302
O IA    10.11.0.0 [110/2] via 10.8.2.254, 00:01:22, Vlan302
```

```
O IA    10.12.0.0 [110/2] via 10.8.2.254, 00:01:22, Vlan302
O IA    10.13.0.0 [110/2] via 10.8.2.254, 00:01:22, Vlan302
O IA    10.20.0.0 [110/2] via 10.8.1.254, 00:02:38, Vlan301
O IA    10.21.0.0 [110/2] via 10.8.1.254, 00:02:38, Vlan301
O IA    10.22.0.0 [110/2] via 10.8.1.254, 00:02:38, Vlan301
O IA    10.23.0.0 [110/2] via 10.8.1.254, 00:02:38, Vlan301
     172.16.0.0/24 is subnetted, 9 subnets
O IA    172.16.0.0 [110/3] via 10.8.2.254, 00:01:22, Vlan302
O IA    172.16.4.0 [110/3] via 10.8.2.254, 00:01:22, Vlan302
O IA    172.16.8.0 [110/3] via 10.8.2.254, 00:01:22, Vlan302
O IA    172.16.12.0 [110/3] via 10.8.2.254, 00:01:22, Vlan302
O IA    172.16.16.0 [110/3] via 10.8.2.254, 00:01:22, Vlan302
O IA    172.16.20.0 [110/3] via 10.8.2.254, 00:01:22, Vlan302
O IA    172.16.24.0 [110/3] via 10.8.2.254, 00:01:22, Vlan302
O IA    172.16.28.0 [110/3] via 10.8.2.254, 00:01:22, Vlan302
O E2    172.16.32.0 [110/20] via 10.8.2.254, 00:01:22, Vlan302
     172.17.0.0/24 is subnetted, 8 subnets
O IA    172.17.0.0 [110/3] via 10.8.1.254, 00:01:22, Vlan301
O IA    172.17.4.0 [110/3] via 10.8.1.254, 00:01:22, Vlan301
O IA    172.17.8.0 [110/3] via 10.8.1.254, 00:01:22, Vlan301
O IA    172.17.12.0 [110/3] via 10.8.1.254, 00:01:22, Vlan301
O IA    172.17.16.0 [110/3] via 10.8.1.254, 00:00:52, Vlan301
O IA    172.17.20.0 [110/3] via 10.8.1.254, 00:00:52, Vlan301
O IA    172.17.24.0 [110/3] via 10.8.1.254, 00:00:52, Vlan301
O IA    172.17.28.0 [110/3] via 10.8.1.254, 00:00:52, Vlan301
     172.20.0.0/24 is subnetted, 2 subnets
O       172.20.1.0 [110/2] via 10.8.0.254, 00:02:18, Vlan300
O       172.20.2.0 [110/2] via 10.8.0.254, 00:02:18, Vlan300
     172.30.0.0/24 is subnetted, 2 subnets
C       172.30.0.0 is directly connected, FastEthernet0/24
O E2    172.30.1.0 [110/20] via 10.8.1.254, 00:02:38, Vlan301
S*      0.0.0.0/0 [1/0] via 172.30.0.254
```

**注意：**

O IA 路由条目为 OSPF 其他区域的路由。

⑥ 在 CORE1 上使用 show ip ospf database 命令查看链路状态数据库：

```
core1#show ip ospf database
        OSPF Router with ID (10.5.0.1) (Process ID 100)
           Router Link States (Area 0)
Link ID         ADV Router      Age         Seq#       Checksum Link count
10.5.0.1        10.5.0.1        313         0x80000007 0x00fd96 3
10.5.0.4        10.5.0.4        313         0x80000006 0x003728 3
10.5.0.2        10.5.0.2        295         0x80000006 0x00f8d1 2
10.5.0.3        10.5.0.3        295         0x80000007 0x00c207 2
           Net Link States (Area 0)
Link ID         ADV Router      Age         Seq#       Checksum
10.8.1.1        10.5.0.1        333         0x80000001 0x007762
10.8.2.1        10.5.0.1        323         0x80000002 0x0023ea
10.8.0.1        10.5.0.1        313         0x80000003 0x00e82b
10.8.3.254      10.5.0.3        295         0x80000001 0x0098bf
```

```
               Summary Net Link States (Area 0)
!!! 类型 3 的 LSA 用于宣告区域间路由
Link ID         ADV Router      Age         Seq#       Checksum
10.20.0.0       10.5.0.2        342         0x80000001 0x00919d
10.21.0.0       10.5.0.2        342         0x80000002 0x0083a9
10.22.0.0       10.5.0.2        342         0x80000003 0x0075b5
10.23.0.0       10.5.0.2        342         0x80000004 0x0067c1
10.10.0.0       10.5.0.3        260         0x80000001 0x000434
10.11.0.0       10.5.0.3        260         0x80000002 0x00f540
10.12.0.0       10.5.0.3        260         0x80000003 0x00e74c
10.13.0.0       10.5.0.3        260         0x80000004 0x00d958
172.16.0.0      10.5.0.3        260         0x80000006 0x007910
172.16.4.0      10.5.0.3        260         0x80000007 0x004b39
172.16.8.0      10.5.0.3        260         0x80000008 0x001d62
172.16.12.0     10.5.0.3        260         0x80000009 0x00ee8b
172.16.16.0     10.5.0.3        260         0x8000000a 0x00c0b4
172.16.20.0     10.5.0.3        260         0x8000000b 0x0092dd
172.16.24.0     10.5.0.3        260         0x8000000c 0x006407
172.16.28.0     10.5.0.3        260         0x8000000d 0x003630
172.17.0.0      10.5.0.2        257         0x80000005 0x007515
172.17.4.0      10.5.0.2        257         0x80000006 0x00473e
172.17.8.0      10.5.0.2        257         0x80000007 0x001967
172.17.12.0     10.5.0.2        257         0x80000008 0x00ea90
172.17.16.0     10.5.0.2        233         0x80000009 0x00bcb9
172.17.20.0     10.5.0.2        233         0x8000000a 0x008ee2
172.17.24.0     10.5.0.2        228         0x8000000b 0x00600c
172.17.28.0     10.5.0.2        228         0x8000000c 0x003235
               Summary ASB Link States (Area 0)
!!! 类型 4 的 LSA 用于指出自治系统边界路由器(ASBR)的位置
Link ID         ADV Router      Age         Seq#       Checksum
10.10.0.4       10.5.0.3        260         0x80000005 0x00c569
                Type-5 AS External Link States
!!! 类型 5 的 LSA 用于宣告外部路由
Link ID         ADV Router      Age         Seq#       Checksum Tag
172.30.0.0      10.5.0.1        417         0x80000003 0x000fd4 0
0.0.0.0         10.5.0.1        417         0x80000004 0x008e31 1
172.30.1.0      10.5.0.2        346         0x80000006 0x00f7e6 0
0.0.0.0         10.5.0.2        346         0x80000007 0x008239 1
172.16.32.0     10.10.0.4       283         0x80000003 0x001cad 0
```

 注意：

因为 CORE1 为骨干路由器，而且仅在 area0 中，所以只能看到 area 0 的链路状态数据库。

⑦ 在 CORE2 上使用 show ip ospf database 命令查看链路状态数据库：

```
core2#show ip ospf database
       OSPF Router with ID (10.5.0.2) (Process ID 100)
          Router Link States (Area 0)
Link ID         ADV Router      Age         Seq#       Checksum Link count
10.5.0.1        10.5.0.1        694         0x80000007 0x00fd96 3
10.5.0.4        10.5.0.4        694         0x80000006 0x003728 3
10.5.0.2        10.5.0.2        676         0x80000006 0x00f8d1 2
```

```
10.5.0.3         10.5.0.3         676       0x80000007 0x00c207 2
                 Net Link States (Area 0)
Link ID          ADV Router       Age       Seq#       Checksum
10.8.1.1         10.5.0.1         714       0x80000001 0x007762
10.8.2.1         10.5.0.1         704       0x80000002 0x0023ea
10.8.0.1         10.5.0.1         694       0x80000003 0x00e82b
10.8.3.254       10.5.0.3         676       0x80000001 0x0098bf
                 Summary Net Link States (Area 0)
Link ID          ADV Router       Age       Seq#       Checksum
10.20.0.0        10.5.0.2         723       0x80000001 0x00919d
10.21.0.0        10.5.0.2         723       0x80000002 0x0083a9
10.22.0.0        10.5.0.2         723       0x80000003 0x0075b5
10.23.0.0        10.5.0.2         723       0x80000004 0x0067c1
10.10.0.0        10.5.0.3         641       0x80000001 0x000434
10.11.0.0        10.5.0.3         641       0x80000002 0x00f540
10.12.0.0        10.5.0.3         641       0x80000003 0x00e74c
10.13.0.0        10.5.0.3         641       0x80000004 0x00d958
172.16.0.0       10.5.0.3         641       0x80000006 0x007910
172.16.4.0       10.5.0.3         641       0x80000007 0x004b39
172.16.8.0       10.5.0.3         641       0x80000008 0x001d62
172.16.12.0      10.5.0.3         641       0x80000009 0x00ee8b
172.16.16.0      10.5.0.3         641       0x8000000a 0x00c0b4
172.16.20.0      10.5.0.3         641       0x8000000b 0x0092dd
172.16.24.0      10.5.0.3         641       0x8000000c 0x006407
172.16.28.0      10.5.0.3         641       0x8000000d 0x003630
172.17.0.0       10.5.0.2         638       0x80000005 0x007515
172.17.4.0       10.5.0.2         638       0x80000006 0x00473e
172.17.8.0       10.5.0.2         638       0x80000007 0x001967
172.17.12.0      10.5.0.2         638       0x80000008 0x00ea90
172.17.16.0      10.5.0.2         613       0x80000009 0x00bcb9
172.17.20.0      10.5.0.2         613       0x8000000a 0x008ee2
172.17.24.0      10.5.0.2         608       0x8000000b 0x00600c
172.17.28.0      10.5.0.2         608       0x8000000c 0x003235
                 Summary ASB Link States (Area 0)
Link ID          ADV Router       Age       Seq#       Checksum
10.10.0.4        10.5.0.3         641       0x80000005 0x00c569
                 Router Link States (Area 20)
Link ID          ADV Router       Age       Seq#       Checksum Link count
10.20.0.2        10.20.0.2        644       0x80000005 0x005ec2 3
10.20.0.1        10.20.0.1        644       0x80000006 0x00a68d 3
10.20.0.3        10.20.0.3        624       0x80000006 0x0012f9 3
10.5.0.2         10.5.0.2         614       0x80000008 0x000ee5 4
10.20.0.4        10.20.0.4        614       0x80000006 0x00c730 3
                 Net Link States (Area 20)
Link ID          ADV Router       Age       Seq#       Checksum
10.21.0.1        10.5.0.2         644       0x80000001 0x004e2f
10.20.0.1        10.5.0.2         644       0x80000002 0x00522c
10.22.0.1        10.5.0.2         624       0x80000003 0x0006a7
10.23.0.1        10.5.0.2         614       0x80000004 0x00fdac
                 Summary Net Link States (Area 20)
Link ID          ADV Router       Age       Seq#       Checksum
10.8.1.0         10.5.0.2         669       0x80000001 0x001723
10.8.3.0         10.5.0.2         669       0x80000002 0x00fe38
```

```
10.8.0.0        10.5.0.2        669         0x80000004 0x002611
10.8.2.0        10.5.0.2        669         0x80000005 0x000e26
172.20.1.0      10.5.0.2        669         0x80000006 0x004e36
172.20.2.0      10.5.0.2        669         0x80000007 0x004141
10.10.0.0       10.5.0.2        614         0x80000008 0x00062b
10.11.0.0       10.5.0.2        614         0x80000009 0x00f737
10.12.0.0       10.5.0.2        614         0x8000000a 0x00e943
10.13.0.0       10.5.0.2        614         0x8000000b 0x00db4f
172.16.0.0      10.5.0.2        614         0x8000000c 0x007d06
172.16.4.0      10.5.0.2        614         0x8000000d 0x004f2f
172.16.8.0      10.5.0.2        614         0x8000000e 0x002158
172.16.12.0     10.5.0.2        614         0x8000000f 0x00f281
172.16.16.0     10.5.0.2        614         0x80000010 0x00c4aa
172.16.20.0     10.5.0.2        614         0x80000011 0x0096d3
172.16.24.0     10.5.0.2        614         0x80000012 0x0068fc
172.16.28.0     10.5.0.2        614         0x80000013 0x003a26
             Summary ASB Link States (Area 20)
Link ID         ADV Router      Age         Seq#       Checksum
10.5.0.1        10.5.0.2        669         0x80000003 0x002a10
             Type-5 AS External Link States
Link ID         ADV Router      Age         Seq#       Checksum Tag
172.30.1.0      10.5.0.2        727         0x80000006 0x00f7e6 0
0.0.0.0         10.5.0.2        727         0x80000007 0x008239 1
172.30.0.0      10.5.0.2        798         0x80000003 0x000fd4 0
0.0.0.0         10.5.0.1        798         0x80000004 0x008e31 1
172.16.32.0     10.10.0.4       664         0x80000003 0x001cad 0
172.30.1.0      10.5.0.2        728         0x80000004 0x00fbe4 0
0.0.0.0         10.5.0.2        728         0x80000005 0x008637 1
```

> **注意**
>
> 因为CORE2为骨干路由器，同时又是边界路由器（ABR）、自治系统边界路由器（ABSR），所以只能看到多个area的链路状态数据库。

⑧ 可以尝试按照步骤⑦方法，分析LSA的传输范围。

总结：LSA的传输范围为area内部，area之间的路由更新由ABR来完成。

⑨ OSPF规划表如表3-1所示。

表3-1  OSPF规划表

| 园　　区 | 楼　　宇 | OSPF AREA |
| --- | --- | --- |
| CZIE | 1#公寓 | 10 |
| | 2#公寓 | |
| | 3#公寓 | |
| | 4#公寓 | |
| | 教学楼 | 20 |
| | 实训楼 | |
| | 图书馆 | |
| | 综合楼 | |
| | 主楼（信息中心） | 0 |

（3）路由汇总

① 按照配置脚本配置核心层、汇聚层、接入层设备。

② 通过 show ip route 命令查看汇聚层设备路由。下面列出了 DS_1_GY 的路由表：

```
DS_1_GY#show ip route
     10.0.0.0/24 is subnetted, 12 subnets
O IA    10.8.0.0 [110/3] via 10.10.0.1, 00: 04: 03, Vlan310
O IA    10.8.1.0 [110/3] via 10.10.0.1, 00: 04: 03, Vlan310
O IA    10.8.2.0 [110/2] via 10.10.0.1, 00: 04: 03, Vlan310
O IA    10.8.3.0 [110/2] via 10.10.0.1, 00: 04: 03, Vlan310
C       10.10.0.0 is directly connected, Vlan310
O       10.11.0.0 [110/2] via 10.10.0.1, 00: 04: 03, Vlan310
O       10.12.0.0 [110/2] via 10.10.0.1, 00: 04: 03, Vlan310
O       10.13.0.0 [110/2] via 10.10.0.1, 00: 04: 03, Vlan310
O IA    10.20.0.0 [110/3] via 10.10.0.1, 00: 03: 38, Vlan310
O IA    10.21.0.0 [110/3] via 10.10.0.1, 00: 03: 38, Vlan310
O IA    10.22.0.0 [110/3] via 10.10.0.1, 00: 03: 38, Vlan310
O IA    10.23.0.0 [110/3] via 10.10.0.1, 00: 03: 38, Vlan310
     172.16.0.0/24 is subnetted, 9 subnets
C       172.16.0.0 is directly connected, FastEthernet0/23
C       172.16.4.0 is directly connected, Vlan110
O       172.16.8.0 [110/3] via 10.10.0.1, 00: 03: 53, Vlan310
O       172.16.12.0 [110/3] via 10.10.0.1, 00: 03: 53, Vlan310
O       172.16.16.0 [110/3] via 10.10.0.1, 00: 03: 38, Vlan310
O       172.16.20.0 [110/3] via 10.10.0.1, 00: 03: 38, Vlan310
O       172.16.24.0 [110/3] via 10.10.0.1, 00: 03: 38, Vlan310
O       172.16.28.0 [110/3] via 10.10.0.1, 00: 03: 38, Vlan310
O E2    172.16.32.0 [110/20] via 10.10.0.1, 00: 03: 38, Vlan310
     172.17.0.0/24 is subnetted, 8 subnets
O IA    172.17.0.0 [110/4] via 10.10.0.1, 00: 01: 02, Vlan310
O IA    172.17.4.0 [110/4] via 10.10.0.1, 00: 01: 02, Vlan310
O IA    172.17.8.0 [110/4] via 10.10.0.1, 00: 01: 02, Vlan310
O IA    172.17.12.0 [110/4] via 10.10.0.1, 00: 01: 02, Vlan310
O IA    172.17.16.0 [110/4] via 10.10.0.1, 00: 01: 02, Vlan310
O IA    172.17.20.0 [110/4] via 10.10.0.1, 00: 01: 02, Vlan310
O IA    172.17.24.0 [110/4] via 10.10.0.1, 00: 01: 02, Vlan310
O IA    172.17.28.0 [110/4] via 10.10.0.1, 00: 01: 02, Vlan310
     172.20.0.0/24 is subnetted, 2 subnets
O IA    172.20.1.0 [110/4] via 10.10.0.1, 00: 04: 03, Vlan310
O IA    172.20.2.0 [110/4] via 10.10.0.1, 00: 04: 03, Vlan310
O*E2 0.0.0.0/0 [110/1] via 10.10.0.1, 00: 00: 00, Vlan310
```

③ 很明显，核心层、汇聚层路由信息已经收敛完成，但是路由信息量太大，因为 OSPF 默认情况下不会自动汇总，占用系统资源比较大。考虑将多个子网汇总成：172.16.0.0/16、172.17.0.0/16。

④ 在核心层进行汇总，如 CORE3 交换机：将 172.16.0.0/24……172.16.28.9/24 汇总成：172.16.0.0/16。

⑤ 通常在 AREA 边界，如 CORE 上进行汇总，如 CORE3：

```
Core3(config-router)# area 0 10.8.0.0 255.255.0.0
Core3(config-router)#area 10 172.16.0.0 255.255.0.0
```

**注意**

汇总命令在 Packet tracert 5.3 上无法实现。

**（4）Stub 部署**

在实际网络应用中，汇聚层路由并不需要学习到整个网络路由。过多的路由条目会导致路由表变大，不利于路由查询，同时也不利于网络的收敛。OSPF 网络路由是通过 LSA 来进行通告的，可以限制 LSA 的泛洪范围，并辅以适当的汇总路由，从而有效地减小路由表的大小，便于网络的收敛，同时增强网络的稳定性。

① 在 DS_1_GY 上查看链路状态数据库：

```
DS_1_GY#show ip ospf database
            OSPF Router with ID (10.10.0.1) (Process ID 100)
            Router Link States (Area 10)
Link ID         ADV Router      Age         Seq#       Checksum Link count
10.10.0.1       10.10.0.1       48          0x80000004 0x00fb66 3
10.10.0.2       10.10.0.2       47          0x80000004 0x00212d 3
10.10.0.3       10.10.0.3       47          0x80000004 0x0046f3 3
10.5.0.3        10.5.0.3        47          0x80000008 0x004efd 4
10.10.0.4       10.10.0.4       47          0x80000005 0x006fb3 3
            Net Link States (Area 10)
Link ID         ADV Router      Age         Seq#       Checksum
10.10.0.254     10.10.0.1       48          0x80000001 0x00e22a
10.11.0.254     10.10.0.2       47          0x80000001 0x000a5e
10.12.0.254     10.10.0.3       47          0x80000001 0x000263
10.13.0.254     10.10.0.4       47          0x80000001 0x00f968
            Summary Net Link States (Area 10)
Link ID         ADV Router      Age         Seq#       Checksum
10.8.2.0        10.5.0.3        88          0x80000001 0x000632
10.8.3.0        10.5.0.3        88          0x80000002 0x00f83d
10.8.1.0        10.5.0.3        32          0x80000004 0x001520
10.8.0.0        10.5.0.3        32          0x80000005 0x001e17
10.20.0.0       10.5.0.3        32          0x80000006 0x008b9c
10.21.0.0       10.5.0.3        32          0x80000007 0x007da8
10.22.0.0       10.5.0.3        32          0x80000008 0x006fb4
10.23.0.0       10.5.0.3        32          0x80000009 0x0061c0
172.17.0.0      10.5.0.3        32          0x8000000a 0x006f14
172.17.4.0      10.5.0.3        32          0x8000000b 0x00413d
172.17.8.0      10.5.0.3        32          0x8000000c 0x001366
172.17.12.0     10.5.0.3        32          0x8000000d 0x00e48f
172.17.16.0     10.5.0.3        32          0x8000000e 0x00b6b8
172.17.20.0     10.5.0.3        32          0x8000000f 0x0088e1
172.17.24.0     10.5.0.3        32          0x80000010 0x005a0b
172.17.28.0     10.5.0.3        32          0x80000011 0x002c34
            Summary ASB Link States (Area 10)         !!! LSA 4
Link ID         ADV Router      Age         Seq#       Checksum
10.5.0.1        10.5.0.3        32          0x80000003 0x002415
            Type-5 AS External Link States            !!! LSA 5
Link ID         ADV Router      Age         Seq#       Checksum Tag
172.30.1.0      10.5.0.2        92          0x80000001 0x0002e1 0
```

```
0.0.0.0            10.5.0.2          92         0x80000002 0x008c34 1
172.30.0.0         10.5.0.1          92         0x80000001 0x0013d2 0
0.0.0.0            10.5.0.1          92         0x80000002 0x00922f 1
172.16.32.0        10.10.0.4         91         0x80000001 0x0020ab 0
```

② 在 CORE3 和 DS_1_GY、DS_2_GY、DS_3_GY、DS_4_GY 上配置 area 1 区域为 stub 区域。

③ 在 CORE3 上配置：

```
router ospf 100
area 10 stub
```

④ 在 DS_1_GY、DS_2_GY、DS_3_GY、DS_4_GY 上配置：

```
router ospf 100
 area 10 stub
!!! 配置 area 10 区域为末结区域
```

**注意**

只要是从属于 area 10 区域的 OSPF 路由器，都需要配置此命令。

⑤ 在 CORE3 上将出现以下提示：

```
00:38:13: %OSPF-5-ADJCHG: Process 100, Nbr 10.10.0.1 on Vlan310 from FULL
to DOWN, Neighbor Down: Adjacency forced to reset
00:38:13: %OSPF-5-ADJCHG: Process 100, Nbr 10.10.0.1 on Vlan310 from FULL
to Down: Interface down or detached
```

⑥ 再次查看 DS_1_GY 的链路状态数据库：

```
DS_1_GY#show ip ospf database
        OSPF Router with ID (10.10.0.1) (Process ID 100)
            Router Link States (Area 10)
Link ID         ADV Router       Age        Seq#        Checksum Link count
10.10.0.1       10.10.0.1        180        0x80000004 0x00fb66 3
10.5.0.3        10.5.0.3         180        0x80000006 0x00701d 4
            Net Link States (Area 10)
Link ID         ADV Router       Age        Seq#        Checksum
10.10.0.254     10.10.0.1        180        0x80000001 0x006e56
            Summary Net Link States (Area 10)
Link ID         ADV Router       Age        Seq#        Checksum
0.0.0.0         10.5.0.3         220        0x80000001 0x00fe4d
10.8.2.0        10.5.0.3         215        0x80000002 0x000433
10.8.3.0        10.5.0.3         215        0x80000003 0x00f63e
10.8.1.0        10.5.0.3         165        0x80000004 0x001520
10.8.0.0        10.5.0.3         164        0x80000005 0x001e17
172.20.1.0      10.5.0.3         164        0x80000006 0x00483b
172.20.2.0      10.5.0.3         164        0x80000007 0x003b46
10.20.0.0       10.5.0.3         164        0x80000008 0x00879e
10.21.0.0       10.5.0.3         164        0x80000009 0x0079aa
10.22.0.0       10.5.0.3         164        0x8000000a 0x006bb6
10.23.0.0       10.5.0.3         164        0x8000000b 0x005dc2
172.17.0.0      10.5.0.3         164        0x8000000c 0x006b16
172.17.4.0      10.5.0.3         164        0x8000000d 0x003d3f
172.17.8.0      10.5.0.3         164        0x8000000e 0x000f68
172.17.12.0     10.5.0.3         164        0x8000000f 0x00e091
```

```
172.17.24.0      10.5.0.3         164      0x80000010 0x005a0b
172.17.28.0      10.5.0.3         164      0x80000011 0x002c34
```

> **注意**
> 由于在 CORE3 上配置 area 10 为末结区域，因此 CORE3 必须发送一条默认路由以确保 area 10 区域的路由器通过它访问非 OSPF 的网络。可以发现类型 4 和 5 的 LSA 被拒绝了，从而限制了 LSA 的泛洪范围。

⑦ 再次查看 DS_1_GY 的路由表，确认路由表变化。因为有 CORE3 发送的默认路由，因此 DS_1_GY 也不需要接收类型 3 的 LSA，即无须知晓 ASBR 的位置。

```
DS_1_GY#show ip route
     10.0.0.0/24 is subnetted, 12 subnets
O IA    10.8.0.0 [110/3] via 10.10.0.1, 00:05:49, Vlan310
O IA    10.8.1.0 [110/3] via 10.10.0.1, 00:05:49, Vlan310
O IA    10.8.2.0 [110/2] via 10.10.0.1, 00:06:04, Vlan310
O IA    10.8.3.0 [110/2] via 10.10.0.1, 00:06:04, Vlan310
C       10.10.0.0 is directly connected, Vlan310
O       10.11.0.0 [110/2] via 10.10.0.1, 00:06:04, Vlan310
O       10.12.0.0 [110/2] via 10.10.0.1, 00:06:04, Vlan310
O       10.13.0.0 [110/2] via 10.10.0.1, 00:06:04, Vlan310
O IA    10.20.0.0 [110/3] via 10.10.0.1, 00:05:49, Vlan310
O IA    10.21.0.0 [110/3] via 10.10.0.1, 00:05:49, Vlan310
O IA    10.22.0.0 [110/3] via 10.10.0.1, 00:05:49, Vlan310
O IA    10.23.0.0 [110/3] via 10.10.0.1, 00:05:49, Vlan310
     172.16.0.0/24 is subnetted, 2 subnets
C       172.16.0.0 is directly connected, FastEthernet0/23
C       172.16.4.0 is directly connected, Vlan110
     172.17.0.0/24 is subnetted, 6 subnets
O IA    172.17.0.0 [110/4] via 10.10.0.1, 00:05:49, Vlan310
O IA    172.17.4.0 [110/4] via 10.10.0.1, 00:05:49, Vlan310
O IA    172.17.8.0 [110/4] via 10.10.0.1, 00:05:49, Vlan310
O IA    172.17.12.0 [110/4] via 10.10.0.1, 00:05:49, Vlan310
O IA    172.17.24.0 [110/4] via 10.10.0.1, 00:05:49, Vlan310
O IA    172.17.28.0 [110/4] via 10.10.0.1, 00:05:49, Vlan310
     172.20.0.0/24 is subnetted, 2 subnets
O IA    172.20.1.0 [110/4] via 10.10.0.1, 00:05:49, Vlan310
O IA    172.20.2.0 [110/4] via 10.10.0.1, 00:05:49, Vlan310
O*IA 0.0.0.0/0 [110/2] via 10.10.0.1, 00:06:04, Vlan310
```

总结：
- 末结区域（stub area）拒绝了类型 4 和类型 5 的 LSA。
- Stub 区域配置要求：Stub 区域没有 ASBR，它至少拥有一个 ABR。

（5）Total Stub 部署

通过 Stub 的区域特性配置，已经可以有效减小路由表的大小。但是此时 R1 的路由表并不是最精简的，可以使用 totally stub 区域特性来进一步减小路由表的大小。

① 配置 CORE3：

```
router ospf 100
   no area 10 stub
   area 10 stub no-summary
```

**注意**

使用 no-summary 命令可以拒绝类型 3 的 LSA 泛洪到 area 1 区域。

② 在 DS_1_GY、DS_2_GY、DS_3_GY、DS_4_GY 上配置：

```
router ospf 100
    area 10 stub
!!! 配置area 10 区域为末结区域
```

③ 再次查看 DS_1_GY 的路由表：

```
DS_1_GY#sh ip route
Gateway of last resort is 10.10.0.1 to network 0.0.0.0
     10.0.0.0/24 is subnetted, 4 subnets
C       10.10.0.0 is directly connected, Vlan310
O       10.11.0.0 [110/2] via 10.10.0.1, 00:00:12, Vlan310
O       10.12.0.0 [110/2] via 10.10.0.1, 00:00:12, Vlan310
O       10.13.0.0 [110/2] via 10.10.0.1, 00:00:12, Vlan310
     172.16.0.0/24 is subnetted, 2 subnets
C       172.16.0.0 is directly connected, FastEthernet0/23
C       172.16.4.0 is directly connected, Vlan110
O*IA 0.0.0.0/0 [110/2] via 10.10.0.1, 00:00:12, Vlan310
```

 **注意**

通过配置完全末结特性，现在 DS_1_GY 只剩下默认路由。

④ 查看 DS_1_GY 的链路状态数据库。现在 DS_1_GY 的链路状态数据库，仅有类型 1 和经过汇总的类型 3 的 LSA，而其他的 OSPF 区域 LSA 被禁止了。

```
DS_1_GY#show ip ospf database
          OSPF Router with ID (10.10.0.1) (Process ID 100)
            Router Link States (Area 10)
Link ID       ADV Router      Age         Seq#        Checksum Link count
10.10.0.1     10.10.0.1       113         0x80000004  0x00fb66 3
10.5.0.3      10.5.0.3        113         0x80000006  0x00701d 4
            Net Link States (Area 10)
Link ID       ADV Router      Age         Seq#        Checksum
10.10.0.254   10.10.0.1       113         0x80000001  0x001259
            Summary Net Link States (Area 10)
Link ID       ADV Router      Age         Seq#        Checksum
0.0.0.0       10.5.0.3        148         0x80000001  0x00fe4d
```

⑤ 使用 ping 命令确认路由：

```
DS_1_GY#ping 172.17.20.254
Type escape sequence to abort.
Sending 5, 100-byte ICMP Echos to 172.17.20.254, timeout is 2 seconds:
!!!!!
Success rate is 100 percent (5/5), round-trip min/avg/max = 31/59/81 ms
```

总结：

- 末结区域（stub area）拒绝了类型 4 和类型 5 的 LSA。
- Stub 区域配置要求：Stub 区域没有 ASBR，它至少拥有一个 ABR。

（6）NSSA 部署

由于 area 1 路由违背了 Stub 区域要求，即 Stub 区域不能够有 ASBR 路由器的特性，因此采用 NSSA 的配置方法来减小 area 20 区域汇聚层路由器（如 DS_JX 等）的路由表大小。

① 在 CORE2 上将 area 20 区域配置成 NSSA 区域：

```
Router ospf 100
    area 20 nssa
!!! 指出area 20区域为nssa
```

② 在 DS_JX、DS_SX、DS_TSG、DS_4_ZH 上配置：

```
router ospf 100
    area 20 nssa
!!! 配置area 20区域为nssa
```

③ 在 DS_JX 上查看路由表：

```
DS_JX#sh ip route
     10.0.0.0/24 is subnetted, 12 subnets
O IA    10.8.0.0 [110/3] via 10.20.0.1, 00:00:03, Vlan320
O IA    10.8.1.0 [110/2] via 10.20.0.1, 00:00:23, Vlan320
O IA    10.8.2.0 [110/3] via 10.20.0.1, 00:00:13, Vlan320
O IA    10.8.3.0 [110/2] via 10.20.0.1, 00:00:23, Vlan320
O IA    10.10.0.0 [110/3] via 10.20.0.1, 00:00:13, Vlan320
O IA    10.11.0.0 [110/3] via 10.20.0.1, 00:00:13, Vlan320
O IA    10.12.0.0 [110/3] via 10.20.0.1, 00:00:13, Vlan320
O IA    10.13.0.0 [110/3] via 10.20.0.1, 00:00:13, Vlan320
C       10.20.0.0 is directly connected, Vlan320
O       10.21.0.0 [110/2] via 10.20.0.1, 00:00:23, Vlan320
O       10.22.0.0 [110/2] via 10.20.0.1, 00:00:23, Vlan320
O       10.23.0.0 [110/2] via 10.20.0.1, 00:00:23, Vlan320
     172.16.0.0/24 is subnetted, 2 subnets
O IA    172.16.0.0 [110/4] via 10.20.0.1, 00:00:13, Vlan320
O IA    172.16.4.0 [110/4] via 10.20.0.1, 00:00:13, Vlan320
     172.17.0.0/24 is subnetted, 2 subnets
C       172.17.0.0 is directly connected, Vlan200
C       172.17.4.0 is directly connected, Vlan210
     172.20.0.0/24 is subnetted, 2 subnets
O IA    172.20.1.0 [110/4] via 10.20.0.1, 00:00:03, Vlan320
O IA    172.20.2.0 [110/4] via 10.20.0.1, 00:00:03, Vlan320
     172.30.0.0/24 is subnetted, 1 subnets
O N2    172.30.1.0 [110/20] via 10.20.0.1, 00:00:23, Vlan320
O*E2 0.0.0.0/0 [110/1] via 10.20.0.1, 00:00:23, Vlan320
```

> **注意**
>
> O IA 路由条目为域键路由。
>
> O N2 路由条目为总外部获得的路由。
>
> O*E2 路由条目用于指出如何到达 CORE2 所连接的外网。CORE2 所连接的外部网络路由较多时，这样的做的好外是不言而喻的。

④ 为了进一步简化 area 20 区域的路由器的路由表,我们采用完全次末结区域(total NSSA)特性来配置 area 20。

在 CORE2 上配置:

```
Router ospf 100
    area 20 nssa no-summary
!!!area 20 为完全次末结区域
!!! no-summary 参数指出不要向区域1发送类型3的区域间汇总路由
```

⑤ 再次在 DS_JX 上查看路由表:

```
DS_JX#sh ip route
     10.0.0.0/24 is subnetted, 4 subnets
C       10.20.0.0 is directly connected, Vlan320
O       10.21.0.0 [110/2] via 10.20.0.1, 00:08:19, Vlan320
O       10.22.0.0 [110/2] via 10.20.0.1, 00:08:19, Vlan320
O       10.23.0.0 [110/2] via 10.20.0.1, 00:08:19, Vlan320
     172.17.0.0/24 is subnetted, 2 subnets
C       172.17.0.0 is directly connected, Vlan200
C       172.17.4.0 is directly connected, Vlan210
     172.30.0.0/24 is subnetted, 1 subnets
O N2    172.30.1.0 [110/20] via 10.20.0.1, 00:08:19, Vlan320
O*IA 0.0.0.0/0 [110/2] via 10.20.0.1, 00:00:07, Vlan320
```

到达其他 OSPF 区域和非 R2 路由器通告的外网路由均被此条默认替代,从而可以精简有效的路由表条目。

⑥ 再次在 DS_JX 上查看链路状态数据库:

```
DS_JX#show ip ospf database
        OSPF Router with ID (10.20.0.1) (Process ID 100)
            Router Link States (Area 20)
Link ID        ADV Router      Age       Seq#     Checksum Link count
10.20.0.1      10.20.0.1       598       0x80000004 0x0096a1 3
10.5.0.2       10.5.0.2        597       0x80000007 0x001149 4
            Net Link States (Area 20)
Link ID        ADV Router      Age       Seq#     Checksum
10.20.0.254    10.20.0.1       598       0x80000001 0x00b6eb
            Summary Net Link States (Area 20)
Link ID        ADV Router      Age       Seq#     Checksum
0.0.0.0        10.5.0.2        110       0x80000010 0x00e657
            Type-7 AS External Link States (Area 20)
Link ID        ADV Router      Age       Seq#     Checksum Tag
172.30.1.0     10.5.0.2        110       0x80000002 0x009e1b 0
            Type-5 AS External Link States
Link ID        ADV Router      Age       Seq#     Checksum Tag
0.0.0.0        10.5.0.2        642       0x80000001 0x008e33 1
```

因为完全次末结区域与完全末结区域类似的是:丢弃类型 3 和类型 4 以及类型 5 的 LSA,所以此处将默认路由转发类型 7,以便进行通告。

从 CORE2 路由器收到类型 7 的 LSA,此类型的 LSA 再送达区域边界路由器(ABR),并将其转换为类型 5 的 LSA 转发给其他区域路由。

总结:在本项目中,可以设计 area 10 为 total stub,area 20 为 total NSSA,以便优化路由。

### 工程化操作

① 安装网络设备，保证物理联通。

② 创建核心层设备配置脚本。新建文本文件，将各个设备配置命令复制、粘贴到其中，保存文件名为：设备名_OSPF_CFG.txt。

创建脚本文件为 CORE1_OSPF_CFG.txt，内容如下：

```
##################     core1- ospf-cfg     ##################
<省略 VLAN 部分>
configure terminal
ip route 0.0.0.0 0.0.0.0 172.30.0.254
router ospf 100
 router-id 10.5.0.1
 redistribute connected subnets
 network 10.8.0.0 0.0.0.255 area 0
 network 10.8.1.0 0.0.0.255 area 0
 network 10.8.2.0 0.0.0.255 area 0
 default-information originate
exit
```

创建脚本文件为 CORE2_OSPF_CFG.txt，内容如下：

```
##################     core2- ospf-cfg     ##################
<省略 VLAN 部分>
ip route 0.0.0.0 0.0.0.0 172.30.1.254
router ospf 100
 router-id 10.5.0.2
 redistribute connected subnets
 network 10.8.1.0 0.0.0.255 area 0
 network 10.8.3.0 0.0.0.255 area 0
 network 10.20.0.0 0.0.0.255 area 20
 network 10.21.0.0 0.0.0.255 area 20
 network 10.22.0.0 0.0.0.255 area 20
 network 10.23.0.0 0.0.0.255 area 20
 area 0 10.8.0.0 255.255.0.0
 area 10 172.17.0.0 255.255.0.0
 area 20 nssa
 default-information originate
exit
```

配置中：粗体部分为路由汇总配置；斜体部分为 NSSA 部署配置。

创建脚本文件为 CORE3_OSPF_CFG.txt，内容如下：

```
##################     core3- ospf-cfg     ##################
<省略 VLAN 部分>
router ospf 100
 router-id 10.5.0.3
 redistribute connected subnets
 network 10.8.2.0 0.0.0.255 area 0
 network 10.8.3.0 0.0.0.255 area 0
 network 10.10.0.0 0.0.0.255 area 10
 network 10.11.0.0 0.0.0.255 area 10
```

```
 network 10.12.0.0 0.0.0.255 area 10
 network 10.13.0.0 0.0.0.255 area 10
 area 0 10.8.0.0 255.255.0.0
 area 10 172.16.0.0 255.255.0.0
 area 10 stub
 exit
```

配置中：粗体部分为路由汇总配置；斜体部分为 Stub 部署配置。

创建脚本文件为 CORE4_OSPF_CFG.txt，内容如下：

```
################## core4- ospf-cfg ##################
<省略 VLAN 部分>
router ospf 100
 router-id 10.5.0.4
 redistribute connected subnets
 network 172.20.0.0 0.0.255.255 area 0
 network 10.8.0.0 0.0.0.255 area 0
exit
```

③ 创建汇聚层设备配置脚本。新建文本文件，将各个设备配置命令复制、粘贴到其中，保存文件名为：设备名_OSPF_CFG.txt。

DS_1_GY 脚本文件为 DS_1_GY_OSPF_CFG.txt，内容如下：

```
################## ds-1-gy- ospf-cfg ##################
<省略 VLAN 部分>
router ospf 100
 router-id 10.10.0.1
 redistribute connected subnets
 network 10.10.0.0 0.0.0.255 area 10
 network 172.16.0.0 0.0.0.255 area 10
 network 172.16.4.0 0.0.0.255 area 10
 area 10 stub no-summary
exit
```

DS_2_GY 脚本文件为 DS_2_GY_OSPF_CFG.txt，内容如下：

```
################## ds-2-gy- ospf-cfg ##################
<省略 VLAN 部分>
router ospf 100
 router-id 10.10.0.1
 redistribute connected subnets
 network 10.11.0.0 0.0.0.255 area 10
 network 172.16.8.0 0.0.0.255 area 10
 network 172.16.12.0 0.0.0.255 area 10
 area 10 stub no-summary
 exit
```

DS_3_GY 脚本文件为 DS_3_GY_OSPF_CFG.txt，内容如下：

```
################## ds-3-gy- ospf-cfg ##################
<省略 VLAN 部分>
router ospf 100
 router-id 10.10.0.1
 redistribute connected subnets
 network 10.12.0.0 0.0.0.255 area 10
 network 172.16.16.0 0.0.0.255 area 10
```

```
 network 172.16.20.0 0.0.0.255 area 10
 area 10 stub no-summary
 exit
```

DS_4_GY 脚本文件为 DS_4_GY_OSPF_CFG.txt，内容如下：

```
##################     ds-4-gy- ospf-cfg     ##################
<省略 VLAN 部分>
router ospf 100
 router-id 10.10.0.1
 redistribute connected subnets
 network 10.13.0.0 0.0.0.255 area 10
 network 172.16.24.0 0.0.0.255 area 10
 network 172.16.28.0 0.0.0.255 area 10
 area 10 stub no-summary
 exit
```

配置中：斜体部分，为 stub 部署配置；公寓楼其他汇聚层交换机配置方法同上。

DS_JX 脚本文件为 DS_JX_OSPF_CFG.txt，内容如下：

```
##################     ds-jx- ospf-cfg     ##################
<省略 VLAN 部分>
router ospf 100
 router-id 10.20.0.1
 redistribute connected subnets
 network 10.20.0.0 0.0.0.255 area 20
 network 172.17.0.0 0.0.0.255 area 20
 network 172.17.4.0 0.0.0.255 area 20
 area 20 nssa
 exit
```

DS_SX 脚本文件为 DS_SX_OSPF_CFG.txt，内容如下：

```
##################     ds-sx- ospf-cfg     ##################
<省略 VLAN 部分>
router ospf 100
 router-id 10.20.0.1
 redistribute connected subnets
 network 10.21.0.0 0.0.0.255 area 20
 network 172.17.8.0 0.0.0.255 area 20
 network 172.17.12.0 0.0.0.255 area 20
 area 20 nssa
 exit
```

DS_TSG 脚本文件为 DS_TSG_OSPF_CFG.txt，内容如下：

```
##################     ds-tsg- ospf-cfg     ##################
<省略 VLAN 部分>
router ospf 100
 router-id 10.20.0.1
 redistribute connected subnets
 network 10.22.0.0 0.0.0.255 area 20
 network 172.17.16.0 0.0.0.255 area 20
 network 172.17.20.0 0.0.0.255 area 20area 20 nssa
 exit
```

DS_ZH 脚本文件为 DS_ZH_OSPF_CFG.txt，内容如下：

```
################    ds-zh- ospf-cfg    ################
<省略 VLAN 部分>
router ospf 100
 router-id 10.20.0.1
 redistribute connected subnets
 network 10.23.0.0 0.0.0.255 area 20
 network 172.17.24.0 0.0.0.255 area 20
 network 172.17.28.0 0.0.0.255 area 20
 area 20 nssa
 exit
```

配置中：斜体部分为 nssa 部署配置；实训楼等其他汇聚层交换机配置方法同上。

④ 执行配置脚本使用 console 口或者远程连接登录设备，打开超级终端或者 SecureCRT 程序，进入特权模式，复制设备中的命令脚本，在超级终端或者 SecureCRT 程序中，右击选择"粘贴"，执行脚本。

⑤ 验证配置：在各设备的特权模式下，使用 show running-config 命令检查当前运行配置；使用 show ip route 命令检查路由；使用 ping 命令检查目标端否可达。

⑥ 其他设备参考 VLAN 规划表参数以及以上方法进行配置和验证，确认无误后，保存设备，并保存设备配置脚本文件。

# 项目 7　访问控制管理

## 项目描述

随着高校校园网络的不断发展和应用需求的不断增多，网络安全问题日益显现出来，这就要求对网络实施精细化的控制。本项目重点分析 ACL 技术的原理，列举了校园网络控制应用实例，说明 ACL 技术在校园网络控制应用中的重要性，并对 ACL 的部署进行设计。

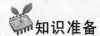

## 知识准备

访问控制列表（access control list，ACL）是路由器和交换机接口的指令列表，用来控制端口进出的数据包。ACL 适用于所有的被路由协议，如 IP、IPX、AppleTalk 等。这张表中包含了匹配关系、条件和查询语句，表只是一个框架结构，其目的是为了对某种访问进行控制。简而言之，ACL 可以过滤网络中的流量，控制访问的一种网络技术手段。

1）ACL 的作用

- ACL 可以限制网络流量、提高网络性能。例如，ACL 可以根据数据包的协议，指定数据包的优先级。
- ACL 提供对通信流量的控制手段。例如，ACL 可以限定或简化路由更新信息的长度，从而限制通过路由器某一网段的通信流量。
- ACL 是提供网络安全访问的基本手段。例如，ACL 允许主机 A 访问人力资源网络，而拒绝主机 B 访问。
- ACL 可以在路由器端口处决定哪种类型的通信流量被转发或被阻塞。例如，用户可以允许 E-mail 通信流量被路由，拒绝所有的 Telnet 通信流量。

例如，某部门要求只能使用 WWW 这个功能，就可以通过 ACL 实现。又例如，为了某部门的保密性，不允许其访问外网，也不允许外网访问它，就可以通过 ACL 实现。

2）ACL 的种类

目前的路由器一般都支持两种类型的访问表：标准访问表和扩展访问表。标准访问表控制基于网络地址的信息流，且只允许过滤源地址。扩展访问表通过网络地址和传输中的数据类型进行信息流控制，允许过滤源地址、目的地址和上层应用数据。表 3-2 列出了路由器所支持的不同访问表的号码范围。

表 3-2 不同访问表的号码范围

| 号码范围 | 访问控制列表种类 | 说　明 |
|---|---|---|
| 1～99 | IP standard access list | 基于 IP 的标准访问控制列表 |
| 100～199 | IP extended access list | 基于 IP 的扩展访问控制列表 |
| 1000～1099 | IPX SAP access list | 基于 IPX 业务通告协议的访问控制列表 |
| 1100～1199 | Extended 48-bit MAC address access list | 扩展的基于 48 位 MAC 地址的访问控制列表 |
| 200～299 | Protocol type-code access list | 基于协议类型码的访问控制列表 |
| 700～799 | 48-bit MAC address access list | 基于 48 位 MAC 地址的访问控制列表 |
| 800～899 | IPX standard access list | 基于 IPX 的标准访问控制列表 |
| 900～999 | IPX extended access list | 基于 IPX 的扩展访问控制列表 |

- 标准 ACL：使用 1～99 以及 800~899 之间的数字作为表号，扩展 ACL 使用 100~199 以及 900~999 之间的数字作为表号。标准 ACL 可以阻止来自某一网络的所有通信流量，或者允许来自某一特定网络的所有通信流量，或者拒绝某一协议簇（比如 IP）的所有通信流量。
- 扩展 ACL：比标准 ACL 提供了更广泛的控制范围。例如，网络管理员如果希望做到"允许外来的 Web 通信流量通过，拒绝外来的 FTP 和 Telnet 等通信流量"，那么，他可以使用扩展 ACL 来达到目的，标准 ACL 不能控制这么精确。

3）标准 ACL 语法

标准 IP 访问表的基本格式为：

access-list [list number][permit|deny][host/any][sourceaddress][wildcard-mask][log]

下面对标准 IP 访问表基本格式中的各项参数进行解释：

- list number——表号范围。标准 IP 访问表的表号从 1 到 99。
- permit|deny——允许或拒绝。关键字 permit 和 deny 用来表示满足访问表项的报文是允许通过接口，还是要过滤掉。permit 表示允许报文通过接口，而 deny 表示匹配标准 IP 访问表源地址的报文要被丢弃掉。
- sourceaddress——源地址。对于标准 IP 访问表，源地址是主机或一组主机的点分十进制表示，如 198.78.46.8。
- host/any——主机匹配。host 和 any 分别用于指定单个主机和所有主机。

host 表示一种精确的匹配，其屏蔽码为 0.0.0.0。例如，假定我们希望允许从 198.78.46.8 来的报文，则语句如下：

access-list 1 permit 198.78.46.8 0.0.0.0

如果采用关键字 host，则也可以用下面的语句来代替：
```
access-list 1 permit host 198.78.46.8
```
也就是说，host 是 0.0.0.0 配符屏蔽码的简写。

与 host 相对照，any 是源地证/目标地址 0.0.0.0/255.255.255.255 的简写。假定我们要拒绝从源地址 198.78.46.8 来的报文，并且允许从其他源地址来的报文，标准 IP 访问表可以使用下面的语句达到这个目的：
```
access-list 1 deny host 198.78.46.8
access-list 1 permit any
```
注意这两条语句的顺序，访问表语句的处理顺序是由上到下的。如果我们将两个语句顺序颠倒，将 permit 语句放在 deny 语句的前面，将不能过滤来自主机地址 198.78.46.8 的报文，因为 permit 语句将允许所有的报文通过。所以说访问表中的语句顺序是很重要的，因为不合理的语句顺序将会在网络中产生安全漏洞，或者使得用户不能很好地利用公司的网络策略。

- wildcard mask——通配符屏蔽码。Cisco 访问表功能所支持的通配符屏蔽码与子网屏蔽码的方式是刚好相反的，也就是说，二进制的 0 表示一个"匹配"条件，二进制的 1 表示一个"不关心"条件。假设组织机构拥有一个 C 类网络 198.78.46.0，若不使用子网，则当配置网络中的每一个工作站时，使用子网屏蔽码 255.255.255.0。在这种情况下，1 表示一个"匹配"条件，而 0 表示一个"不关心"的条件。因为 Cisco 通配符屏蔽码与子网屏蔽码是相反的，所以匹配源网络地址 198.78.46.0 中的所有报文的通配符屏蔽码为：0.0.0.255。

- log——日志记录。log 关键字只在 IOS 版本 11.3 中存在。如果该关键字存在于访问表中，则对那些能够匹配访问表中的 permit 和 deny 语句的报文进行日志记录。日志信息包含访问表号、报文的允许或拒绝、源 IP 地址，以及在显示了第一个匹配以来每 5 分钟间隔内的报文数目。使用 log 关键字，会使控制台日志提供测试和报警两种功能。系统管理员可以使用日志来观察不同活动下的报文匹配情况，从而可以测试不同访问表的设计情况。当其用于报警时，管理员可以察看显示结果，以定位那些多次尝试活动被拒绝的访问表语句。执行一个访问表语句的多次尝试活动被拒绝，很可能表明有潜在的黑客攻击活动。

4）扩展 ACL 语法

顾名思义，扩展 IP 访问表用于扩展报文过滤能力。一个扩展 IP 访问表允许用户根据如下内容过滤报文：源和目的地址、协议、源和目的端口，以及在特定报文字段中允许进行特殊位比较等。扩展 IP 访问表的一般语法格式如下所示：
```
access-list [list number][permit|deny][protocol][sourceaddress]
[source-wildcard-mask][source port][destination address]
[destination-wildcar-mask][destination port][log][option]
```
下面简要介绍各个参数：

- list number——表号范围。扩展 IP 访问表的表号从 100 到 199。
- protocol——协议。protocol 定义了需要被过滤的协议，例如 IP、TCP、UDP、1CMP 等。protocol 选项是很重要的，因为在 TCP/IP 协议栈中的各种协议之间有很密切的关系，如果管理员希望根据特殊协议进行报文过滤，就要指定该协议。

另外，管理员应该注意将相对重要的过滤项放在靠前的位置。如果管理员设置的命令中，允许 IP 地址的语句放在拒绝 TCP 地址的语句前面，则后一个语句根本不起作用。但是如果将这两条语句换一下位置，则在允许该地址上的其他协议的同时，拒绝 TCP 协议。

- source port 和 destination port——源端口号和目的端口号。源端口号可以用几种不同的方法指定。一种是使用一个数字或者使用一个可识别的助记符显式地指定。例如，我们可以使用 80 或者 http 来指定 Web 的超文本传输协议。对于 TCP 和 UDP，读者可以使用操作符<（小于）、>（大于）、=（等于）以及 " "（不等于）来进行设置。

  目的端口号的指定方法与源端口号的指定方法相同。读者可以使用数字、助记符或者使用操作符与数字或助记符相结合的格式来指定一个端口范围。

  下面的实例说明了扩展 IP 访问表中部分关键字的使用方法：
  ```
  access-list 101 permit tcp any host 198.78.46.8 eq smtp
  access-list 101 permit tcp any host 198.78.46.3 eq www
  ```
  第一个语句允许来自任何主机的 TCP 报文到达特定主机 198.78.46.8 的 smtp 服务端口（25）。第二个语句允许任何来自任何主机的 TCP 报文到达指定的主机 198.78.46.3 的 www 或 http 服务端口（80）。

- option 选项。扩展 IP 访问表支持很多选项。其中一个常用的选项是 log，在前面讨论标准访问表时介绍过了。另一个常用的选项是 fistahlishfid，该选项只用于 TCP 协议，并且只在 TCP 通信流的一个方向上来响应由另一端发起的会话。为了实现该功能，使用 established 选项的访问表语句检查每个 TCP 报文，以确定报文的 ACK 或 RST 位是否已设置。

  例如，考虑如下扩展 IP 访问表语句：
  ```
  access-list 101 permit tcp any host 198.78.46.8 established
  ```
  该语句的作用是：只要报文的 ACK 和 RST 位被设置，该访问表语句就允许来自任何源地址的 TCP 报文流到指定的主机 198.78.46.8。这意味着主机 198.78.46.8 此前必须发起 TCP 会话。

- 其他关键字。permit|deny、源地址和通配符屏蔽码、目的地址和通配符屏蔽码以及 host/any 的使用均与标准 IP 访问表中的相同，如表 3-3 所示。

表 3-3　扩展 ACL 关键字

| 关　键　字 | 说　　明 |
| --- | --- |
| any | 对 0.0.0.0、255.255.255.255 的地址和通配符掩码的缩写，用于表示源和目标地址 |
| established | 用于过滤 ACK 或 RST 位是否被置位（只用于 TCP） |
| host | 用于 0.0.0.0 通配符掩码的速写，用于源和目标地址 |
| icmp-type | 用于过滤 ICMP 消息类型，用户也可以指定 ICMP 消息码（0~255） |
| port | 用于定义一个 TCP 或 UDP 端口的十进制号码或名称 |
| protocol | 用于定义要过滤的协议 |
| precedence | 用于优先级名字或数字来过滤 |
| remark | 用于在访问列表中添加文本注释 |
| tos | 用于按数字或名称指定的服务优先级来过滤 |

5）基于命名|编号的 ACL

在标准与扩展访问控制列表中均要使用表号，而在命名访问控制列表中使用一个字母或数字组合的字符串来代替前面所使用的数字。

使用命名访问控制列表可以用来删除某一条特定的控制条目，这样可以让我们在使用过程中方便地进行修改。在使用命名访问控制列表时，要求路由器的IOS为11.2以上的版本，并且不能以同一名字命名多个ACL，不同类型的ACL也不能使用相同的名字。

6）基于时间的ACL

随着网络的发展和用户要求的变化，从IOS 12.0开始，思科（CISCO）路由器新增加了一种基于时间的访问列表。通过它，可以根据一天中的不同时间，或者根据一星期中的不同日期，或二者相结合来控制网络数据包的转发。

这种基于时间的访问列表，就是在原来的标准访问列表和扩展访问列表中，加入有效的时间范围来更合理有效地控制网络。首先定义一个时间范围，然后在原来的各种访问列表的基础上应用它。基于时间的访问列表的设计中，用time-range命令来指定时间范围的名称，然后用absolute命令或者一个或多个periodic命令来具体定义时间范围。

7）ACL执行过程

一个端口执行哪条ACL，这需要按照列表中的条件语句执行顺序来判断。如果一个数据包的报头跟表中某个条件判断语句相匹配，那么后面的语句将被忽略，不再进行检查。

数据包只有在跟第一个判断条件不匹配时，才被交给ACL中的下一个条件判断语句进行比较。如果匹配（假设为允许发送），则不管是第一条还是最后一条语句，数据都会立即发送到目的接口。如果所有的ACL判断语句都检测完毕，仍没有匹配的语句出口，则该数据包将视为被拒绝而被丢弃。这里要注意，ACL不能对本路由器产生的数据包进行控制。

8）定义ACL规范

① ACL的列表号指出了是哪种协议的ACL。各种协议有自己的ACL，而每个协议的ACL又分为标准ACL和扩展ACL，这些ACL是通过ACL列表号区别的。如果在使用一种访问ACL时用错了列表号，那么就会出错误。

② 一个ACL的配置是每协议、每接口、每方向的。路由器的一个接口上每一种协议可以配置进方向和出方向两个ACL。也就是说，如果路由器上启用了IP和IPX两种协议栈，那么路由器的一个接口上可以配置IP、IPX两种协议，每种协议进出两个方向，共四个ACL。

③ ACL的语句顺序决定了对数据包的控制顺序。在ACL中各描述语句的放置顺序是很重要的。当路由器决定某一数据包是被转发还是被阻塞时，会按照各项描述语句在ACL中的顺序，根据各描述语句的判断条件，对数据报进行检查，一旦找到了某一匹配条件就结束比较过程，不再检查以后的其他条件判断语句。

④ 最有限制性的语句应该放在ACL语句的首行。把最有限制性的语句放在ACL语句的首行或者语句中靠近前面的位置，把"全部允许"或者"全部拒绝"这样的语句放在末行或接近末行，可以防止出现诸如本该拒绝（放过）的数据包被放过（拒绝）的情况。

⑤ 新的表项只能被添加到ACL的末尾，这意味着不可能改变已有访问控制列表的功能。如果必须改变，只有先删除已存在的ACL，然后创建一个新ACL，将新ACL应用到相应的接口上。

⑥ 在将ACL应用到接口之前，一定要先建立ACL。首先在全局模式下建立ACL，然后把它应用在接口的出方向或进方向上。在接口上应用一个不存在的ACL是不可能的。

⑦ ACL语句不能被逐条删除，只能一次性删除整个ACL。

⑧ 在 ACL 的最后,有一条隐含的"全部拒绝"的命令,所以在 ACL 里一定至少有一条"允许"的语句。

⑨ ACL 只能过滤穿过路由器的数据流量,不能过滤由本路由器上发出的数据包。

⑩ 在路由器选择进行以前,应用在接口进入方向的 ACL 起作用;在路由器选择决定以后,应用在接口离开方向的 ACL 起作用。

9) ACL 部署

ACL 通过过滤数据包并且丢弃不希望抵达目的地的数据包来控制通信流量。然而,网络能否有效地减少不必要的通信流量,还要取决于网络管理员把 ACL 放置在哪个地方。

假设在一个运行 TCP/IP 协议的网络环境中,网络只想拒绝从 Router A 的 F0 接口连接的网络到 Router D 的 E1 接口连接的网络的访问,即禁止从网络 1 到网络 2 的访问。

根据减少不必要通信流量的通行准则,网管员应该尽可能地把 ACL 放置在靠近被拒绝的通信流量的来源处,即 Router A 上。如果网管员使用标准 ACL 来进行网络流量限制,因为标准 ACL 只能检查源 IP 地址,所以实际执行情况为:凡是检查到源 IP 地址和网络 1 匹配的数据包将会被丢掉,即网络 1 到网络 2、网络 3 和网络 4 的访问都将被禁止。由此可见,这个 ACL 控制方法不能达到网管员的目的。同理,将 ACL 放在 Router B 和 Router C 上也存在同样的问题。只有将 ACL 放在连接目标网络的 Router D 上(E0 接口),网络才能准确实现网管员的目标。由此可以得出一个结论:标准 ACL 要尽量靠近目的端。

网管员如果使用扩展 ACL 来进行上述控制,则完全可以把 ACL 放在 Router A 上,因为扩展 ACL 能控制源地址(网络 1),也能控制目的地址(网络 2),这样从网络 1 到网络 2 访问的数据包在 Router A 上就被丢弃,不会传到 Router B、Router C 和 Router D,从而减少不必要的网络流量。因此,我们可以得出另一个结论:扩展 ACL 要尽量靠近源端。

## 项目实施

1) 标准 ACL 部署

由于 VOD 点播系统为教学科研信息,校园网建设初期,因为没有对访问进行控制,宿舍区很多用户同时访问 VOD 点播系统,导致校园网流量压力很大。根据分析,做出以下调整,校园网用户阻止宿舍区对 VOD 点播系统的访问,其他的都正常使用。

① 定义 ACL:定义标准 ACL 编号为 1,在 CORE3 上部署。

```
CORE3 (config)#access-list 1 deny 172.16.0.0 0.0.0.255
CORE3 (config)#access-list 1 deny 172.16.4.0 0.0.0.255
CORE3 (config)#access-list 1 deny 172.16.8.0 0.0.0.255
CORE3 (config)#access-list 1 deny 172.16.16.0 0.0.0.255
CORE3 (config)#access-list 1 deny 172.16.20.0 0.0.0.255
CORE3 (config)#access-list 1 deny 172.16.24.0 0.0.0.255
CORE3 (config)#access-list 1 deny 172.16.28.0 0.0.0.255
```

② 在接口上应用 ACL:在 CORE3 上检查。

```
CORE3 (config)# interface vlan 302
CORE3 (config)#ip access-group 1 out
CORE3 (config)#exit
CORE3 (config)# interface vlan 303
CORE3 (config)#ip access-group 1 out
```

问题 1：

在 PC0 进行测试，ping VOD 服务器，结果是无法连接。同样，ping CORE3 以外的所有的 IP 时都无法 ping 通！！！

③ 验证配置：在 CORE3 上使用 show ip access-lists 1 命令查看刚刚所配置的 ACL1。

```
Standard IP access list 1
    deny 172.16.0.0 0.0.0.255
    deny 172.16.4.0 0.0.0.255
    deny 172.16.8.0 0.0.0.255
    deny 172.16.16.0 0.0.0.255
    deny 172.16.20.0 0.0.0.255
    deny 172.16.24.0 0.0.0.255
    deny 172.16.28.0 0.0.0.255
```

 注意

在标准 ACL 中，默认隐含最后一条语句：deny any。因为 ACL1 中确实阻止了不该通过的浏览，但是未打开允许的流量，所以导致结果是：所有流量都无法通过。

④ 需要在 ACL1 中在添加一条命令：

CORE3 (config)#access-list 1 permit any

⑤ 当完成配置后，发现宿舍区无法访问 CORE3 以外的所有网络，请分析原因！！！

问题 2：

试想，如果在 CORE3 上未设置 ACL，而在 CORE4 上设置 ACL1，规则相同，在连接 VOD 的接口的 OUT 方向上检查，结果有什么不同？

⑥ 再分析 ACL 的匹配规则，对通配符掩码 0 位进行检查，ACL 中 7 条 deny 语句可以概括成：deny 172.16.0.0.255.255，这样可以简化 ACL 配置工作量。

注意

某设备上配置的 ACL，只能在本地端口应用，而不能在其他设备上调用。

标准访问控制列表应用在离目标段近的位置。

2）扩展 ACL 部署

要求是，为保证 WEB 服务器的安全，只允许校园网内用户访问 WEB 服务（80 端口），其他的拒绝。

① 定义 ACL：在 BR 上，创建标号为 100 的 ACL。

BR(config)#access-list 100 permit tcp any host 172.40.0.1 eq 80

② 在 BR 的 F8/0 口上应用：

```
BR(config)#interface fastEthernet 8/0
BR(config-if)#ip access-group 100 out
```

③ 在拓扑中任何一个 PC 上都可以使用 HTTP 访问，而 ping 172.40.0.1 会发现无法通信。

④ 试想一下，如果在各个楼宇汇聚层交换机配置 HTTP 流量过滤跟在 BR 上配置有什么区别！！！

⑤ 通常在设备上考虑安全，会封锁一些不安全端口：

```
deny    tcp any any eq 139
deny    tcp any any eq 137
```

```
deny    tcp any any eq 135
deny    tcp any any eq 445
deny    tcp any any eq 593
deny    tcp any any eq 1434
deny    tcp any any eq 2500
deny    tcp any any eq 4444
deny    tcp any any eq 5800
deny    tcp any any eq 5900
deny    tcp any any eq 6346
deny    tcp any any eq 6667
deny    tcp any any eq 9393
deny    udp any any eq netbios-ss
deny    udp any any eq 135
deny    udp any any eq 445
deny    udp any any eq 593
deny    udp any any eq 1434
deny    udp any any eq tftp
deny    udp any any eq netbios-dgm
permit  ip any any
```

总结：
- 扩展 ACL 通常放在离源端比较近的位置。
- 对于老版的 IOS，在已配置的 ACL 中无法插入或修改语句，通常是把 ACL 在"记事本"中写好之后，把原有的 ACL 删除，再创建新的，并应用到端口。新版的 IOS 可以插入语句，读者可以在真实设备上尝试（Packet Tracer 无法实现）。

### 工程化操作

① 初始化设备，保证三层路由联通。

② 创建配置脚本。新建文本文件，将各个设备配置命令复制、粘贴到其中，保存文件名为：设备名_ ACL_CFG.txt；

创建脚本文件为 CORE3_ACL_CFG.txt，内容如下：

```
##################    core3-acl-cfg    ##################
configure terminal
access-list 1 deny 172.16.0.0 0.0.0.255
access-list 1 deny 172.16.4.0 0.0.0.255
access-list 1 deny 172.16.8.0 0.0.0.255
access-list 1 deny 172.16.16.0 0.0.0.255
access-list 1 deny 172.16.20.0 0.0.0.255
access-list 1 deny 172.16.24.0 0.0.0.255
access-list 1 deny 172.16.28.0 0.0.0.255
interface vlan 302
ip access-group 1 out
exit
interface vlan 303
ip access-group 1 out
exit
```

创建脚本文件为 BR_ACL_CFG.txt，内容如下：

```
##################    br-acl-cfg    ##################
```

```
configure terminal
ip access-list extended enti-vir
deny    tcp any any eq 139
deny    tcp any any eq 137
deny    tcp any any eq 135
deny    tcp any any eq 445
deny    tcp any any eq 593
deny    tcp any any eq 1434
deny    tcp any any eq 2500
deny    tcp any any eq 4444
deny    tcp any any eq 5800
deny    tcp any any eq 5900
deny    tcp any any eq 6346
deny    tcp any any eq 6667
deny    tcp any any eq 9393
deny    udp any any eq netbios-ss
deny    udp any any eq 135
deny    udp any any eq 445
deny    udp any any eq 593
deny    udp any any eq 1434
deny    udp any any eq tftp
deny    udp any any eq netbios-dgm
permit ip any any
exit
interface fastethernet 4/0
ip access-group enti-vir out
exit
interface fastethernet 5/0
ip access-group enti-vir out
exit
interface fastethernet 8/0
ip access-group enti-vir out
exit
```

配置中防病毒 ACL 使用了基于命名的方式，应用方法与基于编号的相同。

③ 执行配置脚本：使用 console 口或者远程连接登录设备，打开超级终端或者 SecureCRT 程序，进入特权模式，复制设备中的命令脚本，在超级终端或者 SecureCRT 程序中，右击选择"粘贴"，执行脚本。

④ 验证配置：在各设备的特权模式下，使用 show running-config 命令检查当前运行配置；使用 show ip access-list 命令检查 ACL。

⑤ 在相应网段进行测试，验证 ACL 实际效果，确认无误后，保存设备，并保存设备配置脚本文件。

## 项目 8  内网安全

### 项目描述

图 1-3 为校园网拓扑图，为加强校园网安全管理，请对下列环结进行部署：
- 接入交换机准入。
- 日志服务器。

- 时间服务器。
- 用户级别权限。

### 知识准备

交换机端口安全功能，是指针对交换机的端口进行安全属性的配置，从而控制用户的安全接入。交换机端口安全主要有两类：一是限制交换机端口的最大连接数，二是针对交换机端口进行 MAC 地址、IP 地址的绑定。

限制交换机端口的最大连接数可以控制交换机端口下连的主机数，并防止用户进行恶意的 ARP 欺骗。交换机端口的地址绑定，可以针对 IP 地址、MAC 地址、IP + MAC 进行，可以实现对用户进行严格的控制，保证用户的安全接入和防止常见的内网的网络攻击，如 ARP 欺骗、IP 和 MAC 地址欺骗、IP 地址攻击等。

配置了交换机的端口安全功能后，当实际应用超出配置的要求时，将产生一个安全违例，产生安全违例的处理方式有 3 种。

① protect：当安全地址个数满后，安全端口将丢弃未知地址（不是该端口的安全地址中的任何一个）的包。

② restrict：当违例产生时，将发送一个 Trap 通知。

③ shutdown：当违例产生时，将关闭端口并发送一个 Trap 通知。

### 项目实施

1）终端用户准入

校园网由于楼宇、终端设备众多，在保证合法终端使用网络的同时，防止非法用户的接入显得尤为重要，可以使用交换机端口安全（port-security）技术来解决。

① 在 PC0 查看 IP 地址，检查是否能成功获取 DHCP 地址。

② 在 AS_1_GY_1 交换机上，当前仅有 PC0 连接到端口 F0/1，现有另一个用户的笔记本想接入校园网（未通过网络管理员认可），该读者断开 PC0 与 F0/1 口的连接，连接一个交换机 switch1 到 F0/1 口，再通过交换机 switch1 连接 PC0 和 laptop0，如图 3-13 所示。再检查 laptop0 的 IP 信息，如图 3-14 所示，发现也能成功获取 IP 地址信息。

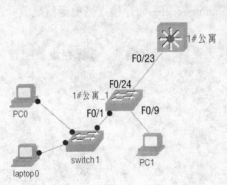

图 3-13 非法交换机接入

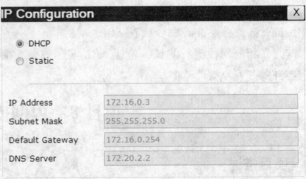

图 3-14 成功获取 IP 地址信息

③ 在 AS_1_GY_1 交换机上使用 show mac-address-table 命令查看 CAM 表：

| Vlan | Mac Address | Type | Ports |
|------|-------------|---------|-------|
| 1 | 0001.63bc.99b2 | DYNAMIC | Fa0/1 |
| 1 | 0060.3e2b.7eed | DYNAMIC | Fa0/1 |
| 1 | 0060.5c44.dec1 | DYNAMIC | Fa0/1 |

在 F0/1 口连接了三个设备：PC0（0060.5c44.dec1）、laptop0（0060.3e2b.7eed）、switch1（0001.63bc.99b2）。

④ 很明显，当前的网络无法识别合法与非法终端身份信息，需要对交换机端口安全进行设置，以防止非法终端接入，在二层区分终端主要通过 MAC 地址。

⑤ 在 AS_1_GY_1 交换机上配置如下：

```
AS_1_GY_1(config)#interface fastethernet 0/1
AS_1_GY_1(config-if)#switchport mode access
AS_1_GY_1(config-if)#switchport port-security
```

F0/1 口被 shutdown，PC0、laptop0 均与接入交换机断开。因为端口安全启用之后默认设置为：连接 MAC 数最大为 1，违例执行动作是 shutdown，当端口因为违例而被关闭后，在全局配置模式下使用命令 errdisable recovery 来将接口从错误状态中恢复过来。或者先使用 shutdown 命令关闭，再使用 no shutdown 命令开启。

⑥ 详细配置可参考表 3-4。

表 3-4 端口安全配置

| 配 置 | 备 注 |
|------|------|
| AS_1_GY_1(config)#interface fastethernet 0/1<br>AS_1_GY_1(config-if)#switchport mode access<br>AS_1_GY_1(config-if)#switchport port-security | 进入 F0/1 口，安全端口必须配置为 access 模式 |
| AS_1_GY_1(config-if)#switchport port-security mac-address 0060.5c44.dec1<br>AS_1_GY_1(config-if)#switchport port-security maximum 1<br>AS_1_GY_1(config-if)#switchport port-security violation protect | 绑定合法 mac 地址<br>设置 mac 地址最大数<br>设置违例方式，思科交换机默认为 shutdown |

⑦ 按照以上配置，当 laptop0 连接到 AS_1_GY_1 交换机，再次验证是否能获取 IP 地址。通过 show port-security、show port-security interface fastethernet 0/1 命令验证。

```
AS_1_GY_1# show port-security:
Secure Port  MaxSecureAddr  CurrentAddr  SecurityViolation  Security Action
             (Count)        (Count)      (Count)
-----------------------------------------------------------------------
  Fa0/1          1              1             1              Shutdown
AS_1_GY_1#show port-security interface fastEthernet 0/1:
Port Security              : Enabled
Port Status                : Secure-shutdown
Violation Mode             : Shutdown
```

```
Aging Time                    : 0 mins
Aging Type                    : Absolute
SecureStatic Address Aging    : Disabled
Maximum MAC Addresses         : 1
Total MAC Addresses           : 1
Configured MAC Addresses      : 1
Sticky MAC Addresses          : 0
Last Source Address:Vlan      : 0001.63BC.99B2:1
Security Violation Count      : 1
```
违例方式申请分析：protect 与 shutdown 应用在什么场合更合理？

> **注意**
> Packet Tracer 5.3 中只支持违例方式：shutdown。为保证网络安全，未使用的交换机端口都应该由管理员手工关闭（shutdown）。

2）日志服务器

对于网络的维护者而言，日志是平时工作中必不可少的一部分，它可以帮助我们分析设备是否正常、网络是否健康，所以任何设备或系统都应该建立完整的日志系统，这样就可以面对任何问题并及时解决问题了。路由器作为重要的网络设备，它的安全性至关重要，由于路由器内存容量有限，一些日志信息存储后掉电就会丢失，所以不能寄希望于将日志保存在路由器上。建立日志服务器，用其记录路由器的运行状况，保存日志记录信息，从而可以帮助我们进行故障定位、故障排除和网络安全管理。

下面实现在日志路由器上记录 BR 路由器的所有记录。

① 配置 syslog 服务器的 IP 地址为 172.20.2.5，日志服务默认开启。

② 在 BR 上配置启用日志命令：

```
BR(config)#logging on
BR(config)# logging 172.20.2.5
```

在终端上提示：%SYS-6-LOGGINGHOST_STARTSTOP: Logging to host 172.20.2.5 port 514 started – CLI initiated，即日志将在 172.20.2.5 上显示。

③ 在日志服务器上查看记录，如图 3-18 所示。

| Time | HostName | Message |
|---|---|---|
| 1　01 00:00:00.000 | 172.30.0.254 | %SYS-5-CONFIG_I: Configured from console … |
| 2　01 00:00:00.000 | 172.30.1.254 | %SYS-6-LOGGINGHOST_STARTSTOP: Logging to host 172.20.2.5 port 514 started – CLI in… |

图 3-18　查看日志服务器记录

④ 在 BR 上定义日志级别为 7 级查看：

```
BR(config)# logging trap 7
BR(config)#logging source-interface e0
```

> **注意**
> 日志消息的级别有 7 种，可以使用 logging trap * 命令进行设置。
> 0:紧急(Emergencies)；1:告警(Alerts)；2:严重的(Critical)；3:错误(Errors)；4:警告(Warnings)；5:通知(Notifications)；6:信息(Informational)；7:调试(Debugging)。

⑤ 在 BR 的 shoudown F1/0 口，查看日志服务器信息，如图 3-19 所示。

| | Time | HostName | Message |
|---|---|---|---|
| 1 | ?? 01 00:00:00.000 | 172.30.1.254 | %LINEPROTO-5-UPDOWN: Line protocol on Int·· |
| 2 | ?? 01 00:00:00.000 | 172.30.0.254 | at1/0, changed state to administratively do·· |
| 3 | ?? 01 00:00:00.000 | 172.30.1.254 | %SYS-6-LOGGINGHOST_STARTSTOP: Logging to host 172.20.2.5 port 514 started - CLI in·· |
| 4 | ?? 01 00:00:00.000 | 172.30.0.254 | %SYS-5-CONFIG_I: Configured from console ·· |
| 5 | ?? 01 00:00:00.000 | 172.30.1.254 | %SYS-5-CONFIG_I: Configured from console ·· |

图 3-19　日志服务器信息

总结：使用日志服务器，可以方便管理员排网络故障，避免日志存放在本地而设备故障无法查看的情况。

3）时间服务器

NTP 服务器，即 Network Time Protocol（NTP），是用来使计算机时间同步化的一种协议，它可以使计算机对其服务器或设备进行同步化，统一网络中所有设备的时间。

① 在 BR 上配置使用 NTP 服务器时间同步：

```
BR(config)# ntp update-calendar
BR(config)# ntp server 172.20.2.5
```

② 如果 NTP 服务器启用认证，需要配置：

```
BR(config)#ntp server authenticate
BR(config)#ntp server authentication key* md5 **
BR(config)#ntp server trusted-key key*
```

4）用户级别权限

IOS 命令行模式下支持两个级别的接入命令：user EXEC（级别 1）和 privileged EXEC（级别 15）。在 0～15 这 16 个级别中，每个级别下所能操作的命令是不一样的，数值越高，级别越高，所能操作的命令也越多。

① 在 BR 路由器上配置三个级别的用户：

```
BR(config)#username cxz1 privilege 14 password 0 111
BR(config)#username cxz2 privilege 7 password 0 222
BR(config)#username cxz3 privilege 2 password 0 333
```

并配置本地 Console 口登录方式：

```
BR(config)#line con 0
BR(config-line)#login local
```

启用本地数据库，使用用户名、密码登录 Console。

② 退出，重新登录到 BR 上进行验证：

| User Access Verification | |
|---|---|
| Username: cxz1 | 使用用户名 cxz1 登录 |
| Password: | 密码为 111 |
| BR#show privilege | 登录成功并用 sh privilege 查看当前用户的优先级 |
| Current privilege level is 14 | 输出显示为 14 |

③ 不同级别的用户之间可以互相切换，首先来看从高级别用户向低级别用户切换：

```
BR#show privilege
```

| | |
|---|---|
| Current privilege level is 14 | 输出显示为14 |
| BR#show privilege<br>Current privilege level is 7; | |
| BR#enable 2 | 登录到优先级为2的用户 |
| BR#sh privilege<br>Current privilege level is 2 | 输出显示为2 |

通过验证，可以看到从高级别向低级别切换时可以直接切换。

④ 尝试级别从低往高进行切换：

| | |
|---|---|
| BR#sh privilege<br>Current privilege level is 2 | 输出显示为2 |
| BR#en 7<br>no password set | 显示说密码没有设置 |

⑤ 从低级别用户向高级别用户切换时需要密码，这也不难理解，下面来设置切换密码：

| | |
|---|---|
| BR(config)#enable secret level 7 cisco | |

设置到级别7用户的切换密码为cisco（这个密码与cxz2的登录密码无关）。

下面进行验证；

| | |
|---|---|
| BR#sh privilege<br>Current privilege level is 2 | 输出显示为2 |
| BR#en 7<br>Password: | 密码为cisco |
| BR#sh privilege<br>Current privilege level is 7 | 输出显示为7 |

配置密码后，低级别用户可以成功登录到高级别用户。

⑥ 目前三个权限的用户都无法进入全局模式，现在要让他们都能进入，并且只有权限14的用户可以用show命令查看。下面进行配置，并进行验证：

| | |
|---|---|
| BR(config)#privilege exec level 2 config t | 设置用户权限大于或等于2时，可以使用configure t命令 |
| BR(config)#privilege exec level 14 show | 设置用户权限大于或等于14时，可以使用show命令 |
| BR#en 2<br>BR#conf t<br>Enter configuration commands, one per line.  End with CNTL/Z.<br>BR(config)# | 权限为2的用户已经可以进入全局模式了 |
| BR(config)#?<br>Configure commands:<br>  default  Set a command to its defaults<br>  end      Exit from configure mode<br>  exit     Exit from configure mode<br>  help    Description of the interactive help system<br>  no      Negate a command or set its defaults | 进入全局模式后只有几条默认的命令，那是因为其他所有的命令都不在这个权限下 |

| | |
|---|---|
| BR#sh ?<br>% Unrecognized command | 权限 2 下无法使用 show 命令 |
| BR#en 7<br>Password:<br>BR#sh ?<br>% Unrecognized command | 权限 7 下也无法使用 show 命令 |
| BR#en 14<br>Password:<br>BR#sh ?<br> ccess-lists   List access lists<br> djacency    Adjacent nodes<br> liases      Display alias commands<br> rp          ARP table<br> sync        Information on terminal lines used as router interfaces<br> ackup      Backup status<br> uffers     Buffer pool statistics<br> ca         CCA information<br> dapi       CDAPI information | 权限 14 下就可以使用 |

**注意**

如果 2 级用户已经赋予了 show start 命令，比如已有 privilege exec level 2 show start，再配置 privilege exec level 5 show start 就会覆盖原配置，并且 show start 的口令级别提升到 5 级，此时再用 2 级用户登录是无法使用 show start 命令的。同样，如果已经配置 privilege exec level 5 show start，再去配置 privilege exec level 2 show start，则前面的配置会被覆盖。也就是说，对于同样的命令，后配置的级别有效。如果配置了 privilege exec level 5 showstar，5 级以下的用户就不能使用 show star 命令了。

## 工程化操作

① 初始化设备，保证三层路由联通。

② 在接入层交换机上部署端口安全，创建配置脚本。新建文本文件，将各个设备配置命令复制、粘贴到其中，保存文件名为：设备名_ port_s_CFG.txt。

创建脚本文件为 as_1_gy_1_CFG.txt，内容如下：

```
################## as-1-gy-1-port-s-cfg ##################
configure terminal
interface fastethernet 0/1
switchport mode access
switchport port-security
switchport port-security mac-address 0060.5c44.dec1
switchport port-security maximum 1
switchport port-security  violation protect
```

其他接入交换机类似部署。

③ 执行配置脚本：使用 console 口或者远程连接登录设备，打开超级终端或者 SecureCRT 程序，进入特权模式，复制设备中的命令脚本，在超级终端或者 SecureCRT 程序中，右击选择"粘贴"，执行脚本。

④ 验证配置：在各设备的特权模式下，使用 show running-config 命令检查当前运行配置，无误后保存设备配置。

⑤ 在出口 BR 上部署 ntp 和 syslog，创建配置脚本。新建文本文件，将各个设备配置命令复制、粘贴到其中，保存文件名为：设备名_cbac_CFG.txt。

创建脚本文件为 br_ntp_slg_CFG.txt，内容如下：

```
################### br-ntp-slg-cfg ##################
configure terminal
logging trap 7
logging source-interface e0
ntp update-calendar
ntp server 172.20.2.5
```

⑥ 执行配置脚本：使用 console 口或者远程连接登录设备，打开超级终端或者 SecureCRT 程序，进入特权模式，复制设备中的命令脚本，在超级终端或者 SecureCRT 程序中，右击选择"粘贴"，执行脚本。

⑦ 验证配置：在各设备的特权模式下，使用 show running-config 命令检查当前运行配置，无误后保存设备配置。

## 本章训练内容

1. 完成第 8 章中项目 18 中任务 2。
2. 完成第 8 章中项目 19 中任务。

# 第 4 章 校园网应用服务

通过前面的项目,我们已经将校园内网基本框架建成,在校园网平台的基础上更重要的是信息服务应用,如果没有应用,校园网存在的价值将大大减弱。

通过本章项目的实践,可以学会校园网中典型的应用系统的部署与分析。

本章需要完成的项目有:

项目 9——部署 DHCP 系统;

项目 10——部署 DNS 系统。

## 项目 9 部署 DHCP 系统

### 项目描述

校园网规模大,接入终端数量众多,如果采用手工方式配置终端地址,就会带来工作量庞大、容易因配置失误导致 IP 地址冲突、不利于网络改造等一系列问题。另外,师生经常携带笔记本在教学楼、图书馆、实训楼等场所跨区域访问校园网,当从一个区域移动到另一个区域时,就需要修改 IP 地址,这都会给日常使用和管理带来很大的问题。本项目要求使用 DHCP 服务器对校园网终端用户 IP 地址进行统一部署与管理。

通过本项目,读者可以掌握以下技能:

① 能够配置 DHCP 服务器。

② 能够分析 DHCP 工作的几个过程。

③ 能够对 DHCP 攻击进行安全防护。

④ 能够分析并配置 DHCP 中继。

### 知识准备

动态主机配置协议(DHCP)是 RFC1541(由 RFC2131 替代)定义的标准协议,该协议允许服务器向客户端动态分配 IP 地址和配置信息。通常,DHCP 服务器至少给客户端提供以下基本信息:IP 地址、子网掩码、默认网关;它还提供其他信息,如域名服务(DNS)服务器地址和 Windows Internet 命名服务(WINS)服务器地址。DHCP 数据包结构如图 4-1 所示。

| | 8 | 16 | 31 |
|---|---|---|---|
| Operation code | Hardware type | Hardware length | Hop count |
| Transaction ID(4B) ||||
| Number of seconds || Flags ||
| Client IP address ||||
| Your IP address ||||
| Server IP address ||||
| Gateway IP address ||||
| Client Hardware address(16B) ||||
| Server name(64B) ||||
| Boot file name(128B) ||||
| Options ||||

图 4-1　DHCP 数据包结构

1）DHCP 报文格式

下面对各字段进行简要说明，见表 4-1。

表 4-1　DHCP 报文字段说明

| 字段 | 说明 |
|---|---|
| Operation code | 若是 Client 送给 Server 的封包，设为 1，反向为 2 |
| Hardware type | 硬件类别，Ethernet 为 1 |
| Hardware length | 硬件地址长度，Ethernet 为 6 |
| Hop count | 若封包需经过 Router 传送，每站加 1，若在同一网内，为 0 |
| Transaction ID | DHCP REQUEST 时产生的数值，作为 DHCPREPLY 时的依据 |
| Number of seconds | Client 端启动时间（秒） |
| Flags | 从 0 到 15 共 16 位，最左一位为 1 时表示 Server 将以广播方式传送封包给 Client，其余尚未使用 |
| Client IP address | Client 端想继续使用之前取得的 IP 地址，则列于这里 |
| Your IP address | 从 Server 送回 Client 之 DHCP OFFER 与 DHCPACK 封包中，此栏填写分配给 Client 的 IP 地址 |
| Server IP address | 若 Client 需要通过网络开机，从 Server 送出至 DHCP OFFER、DHCPACK、DHCPNACK 封包中，此栏填写开机程序代码所在 Server 的地址 |
| Gateway IP address | 若需跨网域进行 DHCP 发放，此栏为 relay agent 的地址，否则为 0 |
| Client Hardware address | Client 的硬件地址 |
| Server name | Server 的名称字符串，以 0x00 结尾 |
| Boot file name | 若 Client 需要通过网络开机，此栏将指出开机程序名称，稍后以 TFTP 传送 |
| Options | 允许厂商定议选项（Vendor-Specific Area），以提供更多的设定信息（如 Netmask、Gateway、DNS 等）。其长度可变，同时可携带多个选项，每一选项之第一个字节为信息代码，其后一个字节 为该项数据长度，最后为项目内容，具体见表 4-2 |

2）DHCP 状态转换

表 4-2 DHCP 报文类型

| Options 值 | 类别 | 说 明 |
| --- | --- | --- |
| 1 | DISCOVER | Client 开始 DHCP 过程中的第一个请求报文 |
| 2 | OFFER | Server 对 DISCOVER 报文的响应 |
| 3 | REQUEST | Client 对 OFFER 报文的响应 |
| 4 | DECLIENT | 当 Client 发现 Server 分配给它的 IP 地址无法使用，如 IP 地址发生冲突时，将发出此报文让 Server 禁止使用这次分配的 IP 地址 |
| 5 | ACK | Server 对 REQUEST 报文的响应，Client 收到此报文后才真正获得 IP 地址和相关配置信息 |
| 6 | NACK | Server 对 Client 的 REQUEST 报文的拒绝响应，Client 收到此报文后，一般会重新开始 DHCP 过程 |
| 7 | RELEASE | Client 主动释放 IP 地址，当 Server 收到此报文后就可以收回 IP 地址分配给其他的 Client |

DHCP 在发送和接收到相关信息时，状态将会进行转换，转换过程可参考图 4-2。

（1）初始化状态（Initializing State）

当 DHCP 客户端第一次启动时，将进入 Initializing 状态，使用 UDP 67 端口发送 DISCOVER 广播。

（2）选择状态（Selecting State）

服务器对客户端的 DISCOVER 回复 OFFER 信息，其中包含提供的 IP 地址、地址租期，服务器同时锁定该地址，不再分配给其他客户端。

客户端收到一个 OFFER（最快的服务器）并发送 REQUEST 信息选择该服务器，并进入 Requesting 状态。如果客户端没有收到任何 OFFER 信息，将每隔 2 秒进行再尝试，共 4 次，如果没有收到任何 DISCOVER 的回应，客户端将停止，5 分钟后继续尝试。

（3）请求状态（Requesting State）

客户端在收到服务器发送的 ACK 信息前都将处于 Requesting 状态，ACK 信息中服务器会绑定客户端的 MAC 和分配的 IP 地址信息，在收到 ACK 信息后，将进入 Bound 状态。

（4）约束状态（Bound State）

该状态下，客户端在租期内可以一直使用 IP 地址，在超过租期的 50%时，客户端将发送另一个 REQUEST 信息续租地址；Bound 状态下，客户端可以取消地址租用重新进入 Initializing 状态。

（5）续租状态（Renewing State）

当客户端重新计时，返回 Bound 状态，若在超过 87.5%的租期还没有收到 ACK 信息，将进入 Rebounding 状态，否则将维持在该状态。

（6）重绑定（Rebinding State）

当客户端收到 NACK 信息或者超期，将进入 Initializing 状态尝试获取另一个 IP 地址，或者客户端收到 ACK 信息进入 Bound 状态重新计时，否则将维持在该状态。

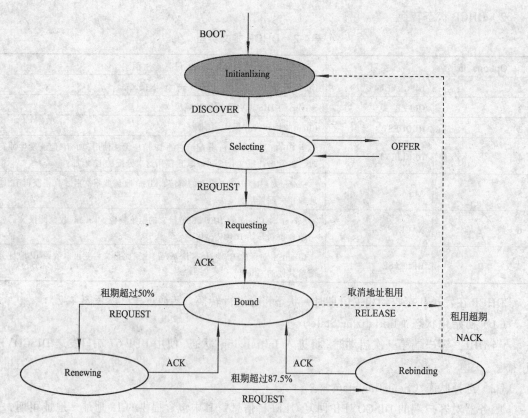

图 4-2 状态转换图

## 项目实施

在服务器群中部署 DHCP 服务器,使用主、备多个服务器来保证 DHCP 的可靠性,由 DHCP 服务器统一管理校园网各个楼宇、部门 IP 地址,使用 DHCP relay、snooping 来优化网络,确保 DHCP 实施效果。

考虑部门隔离与路由优化,设计 DHCP 服务器地址池,如表 4-3 所示,每个子网(VLAN)独立设置一个 Pool(池),其中包含以下参数。

- Pool name:地址池名称。
- Start IP:池中起始 IP 地址。
- Subnet Mask:IP 所对应的子网掩码。
- Max Number:可分配的地址(主机数)最大数。
- Default Gateway:默认网关地址。
- DNS Server:首选 DNS 服务器地址。

表 4-3 全校终端 IP 地址表

| Pool Name | Default Gateway | Start IP | Subnet Mask | DNS Server | Max Number |
|---|---|---|---|---|---|
| Dorm11Pool | 172.16.0.254 | 172.16.0.1 | 255.255.255.0 | 172.20.2.2 | 200 |
| Dorm12Pool | 172.16.4.254 | 172.16.4.1 | 255.255.255.0 | 172.20.2.2 | 200 |

续表

| Pool Name | Default Gateway | Start IP | Subnet Mask | DNS Server | Max Number |
|---|---|---|---|---|---|
| Dorm21Pool | 172.16.8.254 | 172.16.8.1 | 255.255.255.0 | 172.20.2.2 | 200 |
| Dorm22Pool | 172.16.12.254 | 172.16.12.1 | 255.255.255.0 | 172.20.2.2 | 200 |
| Dorm31Pool | 172.16.16.254 | 172.16.16.1 | 255.255.255.0 | 172.20.2.2 | 200 |
| Dorm32Pool | 172.16.20.254 | 172.16.20.1 | 255.255.255.0 | 172.20.2.2 | 200 |
| Dorm41Pool | 172.16.24.254 | 172.16.24.1 | 255.255.255.0 | 172.20.2.2 | 200 |
| Dorm42Pool | 172.16.28.254 | 172.16.28.1 | 255.255.255.0 | 172.20.2.2 | 200 |
| Teaching11Pool | 172.17.0.254 | 172.17.0.1 | 255.255.255.0 | 172.20.2.2 | 200 |
| Teaching12Pool | 172.17.4.254 | 172.17.4.1 | 255.255.255.0 | 172.20.2.2 | 200 |
| Practise11Pool | 172.17.8.254 | 172.17.8.1 | 255.255.255.0 | 172.20.2.2 | 200 |
| Practise12Pool | 172.17.12.254 | 172.17.12.1 | 255.255.255.0 | 172.20.2.2 | 200 |
| Library11Pool | 172.17.16.254 | 172.17.16.1 | 255.255.255.0 | 172.20.2.2 | 200 |
| Library12Pool | 172.17.20.254 | 172.17.20.1 | 255.255.255.0 | 172.20.2.2 | 200 |
| General11Pool | 172.17.24.254 | 172.17.24.1 | 255.255.255.0 | 172.20.2.2 | 200 |
| General12Pool | 172.17.28.254 | 172.17.28.1 | 255.255.255.0 | 172.20.2.2 | 200 |

1）DHCP 服务基础

① 在"DHCP_S-CORE4-CORE1-CORE3-1#公寓"等设备上参照配置全网 OSPF 路由；在项目 6 的基础上，保证 DHCP_S 和 PC 之间畅通，如图 4-3 所示。

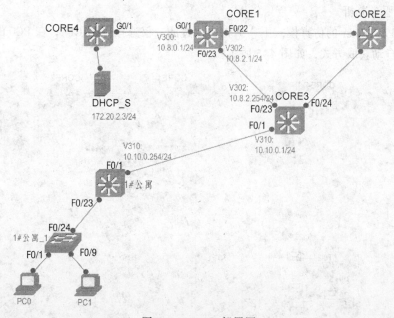

图 4-3 DHCP 部署图

② 在数据中心部署 DHCP Server，IP 地址为 172.20.2.3/24，开启 DHCP 服务功能，为全校终端提供 IP 管理平台，关于 DHCP 服务的配置由于篇幅原因，请参考相关书籍。根据表 4-4 中参数，在 Packet Tracer 5.3 中配置 DHCP 如图 4-4 所示。

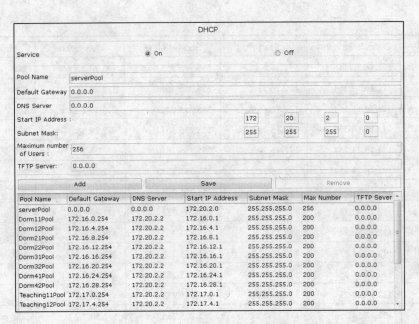

图 4-4 DHCP 服务器地址池配置

③ 在接入终端 PC0 上，配置自动获取 IP、DNS 地址方式，验证 PC0 是否能成功获取 172.16.0.0/24 网段的地址。通过查看，发现 PC0 无法自动获取 IP 地址等信息，结果为无法自动获取 IP 信息。

2）DHCP 服务分析

① 在 Packet Tracer 的仿真模式下，编辑协议过滤，仅选择"DHCP"；在 PC0 的 TCP/IP 配置中再次使用自动获取方式，如图 4-5 所示。

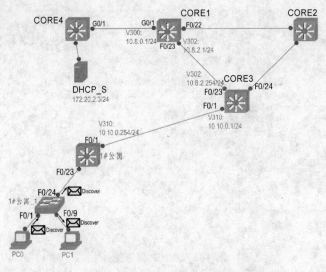

图 4-5 DHCP DISCOVER 包

② 此时为 DHCP 工作的第一个阶段：DISCOVER。分析图 4-6，可以看到：第二层封装中目标地址为 FFFF.FFFFF.FFFF，第三层封装中没有源 IP，目标 IP 为 255.255.255.255，第四层封

装中目标 UDP 端口为 67；PC0 以广播方式（因为 DHCP 服务器的 IP 地址对于客户端来说是未知的）发送 DHCPDISCOVER 信息来查找 DHCP 服务器。

```
At Device: 1#公寓
Source: PC0
Destination: 255.255.255.255
```

| In Layers | Out Layers |
|---|---|
| Layer 7: DHCP Frame Server: 0.0.0.0, Client: 0.0.0.0 | Layer7 |
| Layer6 | Layer6 |
| Layer5 | Layer5 |
| Layer 4: UDP Src Port: 68, Dst Port: 67 | Layer4 |
| Layer 3: IP Header Src. IP: 0.0.0.0, Dest. IP: 255.255.255.255 | Layer 3: IP Header Src. IP: 172.16.0.254, Dest. IP: 172.20.2.3 |
| Layer 2: Ethernet II Header 0060.5C44.DEC1 >> FFFF.FFFF.FFFF | Layer 2: Ethernet II Header 0001.96EA.6DED >> 00D0.D33B.4E63 |
| Layer 1: Port FastEthernet0/23 | Layer 1: Port(s): FastEthernet0/1 |

图 4-6  DISCOVER 封装

③ 报文标准格式参考图 4-7。

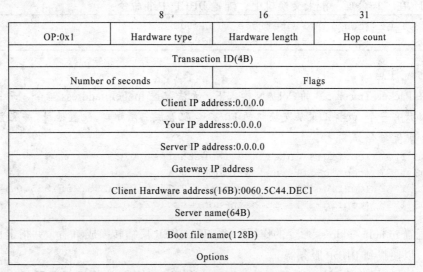

图 4-7  DISCOVER 报文格式

④ 继续捕获数据包，如图 4-8 所示，该包将广播至其他主机和汇聚层设备（其他主机将丢弃），广播到达汇聚层设备时将被隔离，汇聚层交换机将丢弃 DISCOVER 包，请分析原因。

3）DHCP 中继

DISCOVER 因使用广播方式无法穿越汇聚层设备，可以使用 DHCP 中继（relay）方式，将广播转换成单播方式进行，这就需要"1#公寓"汇聚交换机将广播转换成目标地址为 172.20.2.3 的单播，使得 DISCOVER 到达 DHCP_S。

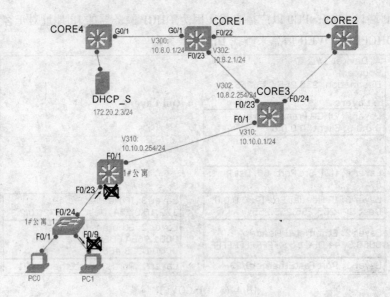

图 4-8 DISCOVER 在汇聚层被丢弃

① 在 "1#公寓" 汇聚交换机上，配置 DHCP 中继命令：

```
DS_1_GY (config)# interface vlan 100
DS_1_GY (config-if)# ip help-address 172.20.2.3
```

### 注意

在 Packet Tracer 5.3 的 VLAN 模式下，无法配置 ip help-address 命令，而且1#公寓接入的VLAN 只有一个，故将汇聚层交换机的 F0/23 口配置成三层端口，配置地址，并配置 ip help-address 地址，配置命令如下：

```
DS_1_GY (config)# interface fastethernet 0/23
DS_1_GY (config-if)# no switchport
DS_1_GY (config-if)# ip address 172.16.0.254 255.255.255.0
DS_1_GY (config-if)# ip help-address 172.20.2.3
```

配置 ip help-address 命令的作用：将 DISCOVER 广播转换成单播，并指定目标端为服务器地址，以便找到 DHCP 服务器。

② 继续捕获数据包，数据封装如图 4-9 所示，汇聚层交换机不再丢弃 DISCOVER 包，而是转换成目标地址为 172.20.2.3 的单播包，而且源 IP 地址为汇聚交换机的 IP 地址。

图 4-9 DHCP relay 封装

③ 继续捕获数据包，DISCOVER 包成功发送到 DHCP_S，如图 4-10 所示。

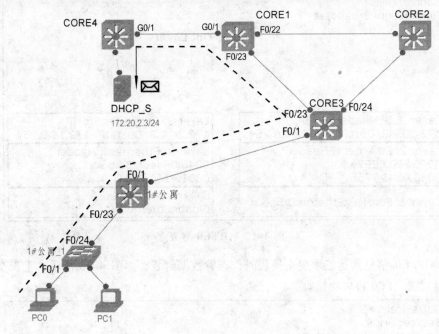

图 4-10 DISCOVER 包发送到 SERVER

④ 进入第二阶段：OFFER，即 DHCP 服务器提供 IP 地址的阶段，在网络中接收到 DHCP DISCOVER 信息的 DHCP 服务器都会做出响应。它从尚未出租的 IP 地址（172.16.0.2）中挑选一个分配给 DHCP 客户端，向其发送一个包含出租的 IP 地址和其他设置的 DHCP OFFER 信息。图 4-11 所示为 OFFER 报文结构，同时包含 DHCP 服务器地址。

| OP:0x2 | Hardware type | Hardware length | Hop count |
|---|---|---|---|
| Transaction ID(4B) ||||
| Number of seconds || Flags ||
| Client IP address:0.0.0.0 ||||
| Your IP address:172.16.0.2 ||||
| Server IP address:172.20.2.3 ||||
| Gateway IP address ||||
| Client Hardware address(16B):0060.5C44.DEC1 ||||
| Server name(64B) ||||
| Boot file name(128B) ||||
| Options ||||

图 4-11 OFFER 报文

⑤ 查看 OFFER 在 CORE4 上的封装，如图 4-12 所示，目标地址为 172.16.0.254，即汇聚层交换机。

```
At Device: Core4
Source: DHCP
Destination: Broadcast
```

| In Layers | Out Layers |
|---|---|
| Layer7 | Layer7 |
| Layer6 | Layer6 |
| Layer5 | Layer5 |
| Layer4 | Layer4 |
| Layer 3: IP Header Src. IP: 172.20.2.3, Dest. IP: 172.16.0.254 | Layer 3: IP Header Src. IP: 172.20.2.3, Dest. IP: 172.16.0.254 |
| Layer 2: Ethernet II Header 0030.F2D7.EE7B >> 0040.0B41.7756 | Layer 2: Ethernet II Header 0040.0B41.7756 >> 0001.426C.371C |
| Layer 1: Port FastEthernet0/5 | Layer 1: Port(s): GigabitEthernet0/1 |

图 4-12 OFFER 封装

⑥ 当 OFFER 信息到达汇聚层交换机时，查看数据封装，如图 4-13 所示，汇聚交换机将把地址广播出去，PC0 将成功接收。

```
At Device: 1#公寓
Source: DHCP
Destination: Broadcast
```

| In Layers | Out Layers |
|---|---|
| Layer7 | Layer7 |
| Layer6 | Layer6 |
| Layer5 | Layer5 |
| Layer4 | Layer4 |
| Layer 3: IP Header Src. IP: 172.20.2.3, Dest. IP: 172.16.0.254 | Layer 3: IP Header Src. IP: 172.16.0.254, Dest. IP: 255.255.255.255 |
| Layer 2: Ethernet II Header 00D0.D33B.4E63 >> 0001.96EA.6DED | Layer 2: Ethernet II Header 0030.A3C6.EA17 >> FFFF.FFFF.FFFF |
| Layer 1: Port FastEthernet0/1 | Layer 1: Port(s): FastEthernet0/23 |

图 4-13 汇聚交换机封装处理

⑦ 在 PC0 的命令行下，使用 ipconfig/all 命令进行查看，IP 地址成功获取。

4）DHCP 防护

DHCP 的第三个阶段：REQUEST，DHCP 客户端选择某台 DHCP 服务器提供的 IP 地址的阶段。如果有多台 DHCP 服务器向 DHCP 客户端发送 DHCP OFFER 信息，则 DHCP 客户端只接收第 1 个发送给它的 DHCP OFFER 信息。然后以广播方式回答一个 DHCP REQUEST 信息，该信息中包含向它所选定的 DHCP 服务器请求 IP 地址的内容。之所以要以广播方式回答，是为了通知所有 DHCP 服务器，它将选择某台 DHCP 服务器所提供的 IP 地址。

① 如图 4-14 所示，如果 PC1 配置 IP 地址为 192.168.0.1，启用 DHCP 服务，提供 192.168.0.0/24 的地址池，PC 将会对 PC1 发送 REQUEST 信息，从而获取 192.168.0.*的非法 IP 地址。

如何解决非法 DHCP 服务器带来的问题，可使用 DHCP snooping 技术实现。

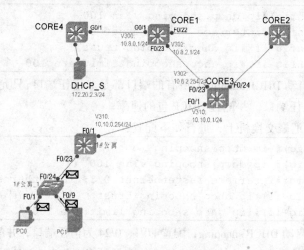

图 4-14 获取非法 IP 地址

通过建立和维护 DHCP snooping 绑定表可以过滤不可信任的 DHCP 信息，这些信息是指来自不信任区域的 DHCP 信息。DHCP snooping 绑定表包含不信任区域的用户 MAC 地址、IP 地址、租用期、VLAN-ID 接口等信息。当交换机开启了 DHCP snooping 后，会对 DHCP 报文进行侦听，并可以从接收到的 DHCP REQUEST 或 DHCP ACK 报文中提取并记录 IP 地址和 MAC 地址信息。另外，DHCP snooping 允许将某个物理端口设置为信任端口或不信任端口。信任端口可以正常接收并转发 DHCP OFFER 报文，而不信任端口会将接收到的 DHCP OFFER 报文丢弃。这样，可以完成交换机对假冒 DHCP Server 的屏蔽作用，确保客户端从合法的 DHCP Server 获取 IP 地址。

② 由 VLAN100 所连接的终端到 DHCP 服务器的链路中，为避免非法 DHCP OFFER 报文，需要确定哪些是信任端口，哪些是非信任端口，图 4-15 中六边形标注的为信任端口。

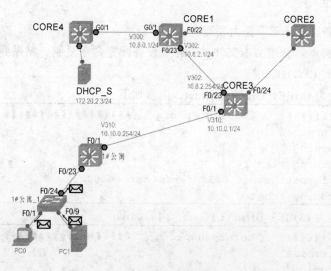

图 4-15 信任端口

③ 在 AS_1_GY_1 接入交换机上，进行如下配置：

```
AS_1_GY_1 (config)# ip dhcp snooping
```

```
AS_1_GY_1 (config)#ip dhcp snooping vlan 100
AS_1_GY_1 (config)#interface fastethernet 0/24
AS_1_GY_1 (config-if)#ip dhcp snooping trust
AS_1_GY_1 (config-if)#ip dhcp snooping limitrate 100
```

在全局模式下,启动 DHCP snooping,所有的端口都是非信任端口,只允许接 PC。在 VLAN100 上,配置 F 0/24 为信任端口,并配置 DHCP 包的速率。

④ 在 DS_1_GY 汇聚交换机上,进行如下配置:

```
DS_1_GY (config)# ip dhcp snooping
DS_1_GY (config)# ip dhcp snooping vlan 100
DS_1_GY (config)#interface fastethernet 0/1, fastethernet 0/23
DS_1_GY (config-if)#ip dhcp snooping trust
DS_1_GY (config-if)#iip dhcp snooping limitrate 100
```

在全局模式下,启动 DHCP snooping,配置 F0/1、0/24 为信任端口,并配置 DHCP 包的速率。

⑤ 在 CORE_1 和 CORE_3 的核心交换机上的配置可参考 DS_1_GY 汇聚交换机配置。

⑥ 在 CORE_4 的核心交换机上,进行配置:

```
CORE_4 (config)# ip dhcp snooping
CORE_4 (config)#ip dhcp snooping vlan 100
CORE_4 (config)#interface gigabitEthernet 0/5, fastethernet 0/23
CORE_4 (config-if)#ip dhcp snooping trust
```

> **注意**
>
> 由于在 Packet Tracert 5.3 上不能完全演示所有状态和报文,读者可以参考 RFC 2131 文档,使用 Windows 操作系统进行捕获数据包分析。

## 工程化操作

① 初始化设备,保证三层路由联通。

② 创建配置脚本。新建文本文件,将各个设备配置命令复制、粘贴到其中,保存文件名为:设备名_ DHCP_CFG.txt。

创建脚本文件为 CORE1_DHCP_CFG.txt,内容如下:

```
##################    core1-dhcp-cfg    ##################
configure terminal
ip dhcp snooping
ip dhcp snooping vlan 100
interface gigabitthernet 0/1, fastethernet 0/23
ip dhcp snooping trust
```

创建脚本文件为 CORE3_DHCP_CFG.txt,内容如下:

```
##################    core3-dhcp-cfg    ##################
configure terminal
ip dhcp snooping
ip dhcp snooping vlan 100
interface fastethernet 0/1, fastethernet 0/23
ip dhcp snooping trust
```

创建脚本文件为 CORE4_DHCP_CFG.txt，内容如下：

```
##################    core4-dhcp-cfg    ##################
configure terminal
ip dhcp snooping
ip dhcp snooping vlan 100
interface gigabitEthernet 0/5, fastethernet 0/23
ip dhcp snooping trust
```

创建脚本文件为 DS_1_GY_DHCP_CFG.txt，内容如下：

```
##################    ds-1-gy-dhcp-cfg    ##################
configure terminal
ip dhcp snooping
ip dhcp snooping vlan 100
interface fastethernet 0/1, fastethernet 0/23
ip dhcp snooping trust
ip dhcp snooping limitrate 100
```

创建脚本文件为 AS_1_GY_1_DHCP_CFG.txt，内容如下：

```
##################    ds-1-gy-dhcp-cfg    ##################
configure terminal
ip dhcp snooping
ip dhcp snooping vlan 100
interface fastethernet 0/4
ip dhcp snooping trust
ip dhcp snooping limitrate 100
```

其他楼宇设备配置可参考以上不同层次交换机的配置。

③ 执行配置脚本：使用 CONSOLE 口或者远程连接登录设备，打开超级终端或者 SecureCRT 程序，进入特权模式，复制设备中的命令脚本，在超级终端或者 SecureCRT 程序中，右击选择"粘贴"，执行脚本。

④ 验证配置：在各设备的特权模式下，使用 show running-config 命令检查当前运行配置。

DS_1_GY 运行配置：

| 配置 | 备注 |
| --- | --- |
| DS_1_GY#show running-config | 查看运行配置 |
| <省略部分><br>interface FastEthernet0/23<br> no switchport<br> ip address 172.16.0.254 255.255.255.0<br> ip helper-address 172.20.2.3<br> duplex auto<br> speed auto<br>!<br><省略部分> | 配置 ip help-address 地址 |
| interface Vlan110<br> ip address 172.16.4.254 255.255.255.0<br>!<br>interface Vlan310<br> ip address 10.10.0.254 255.255.255.0<br>!<br><省略部分> | 配置 VLAN 地址 |

CORE3 运行配置：

| 配置 | 备注 |
|---|---|
| CORE3#show running-config<br>interface fastethernet 0/1<br>switchport access vlan 210<br>ip dhcp snooping limit rate 10<br>ip dhcp snooping trust<br>!<br>interface fastethernet 0/2<br>switchport access vlan 211<br>ip dhcp snooping limit rate 10<br>ip dhcp snooping trust<br>!<br>interface fastethernet 0/3<br>switchport access vlan 212<br>ip dhcp snooping limit rate 10<br>ip dhcp snooping trust<br>!<br>interface fastethernet 0/4<br>switchport access vlan 213<br>ip dhcp snooping limit rate 10<br>ip dhcp snooping trust<br>!<br>&lt;省略部分&gt;<br>interface fastethernet 0/23<br>switchport access vlan 302<br>ip dhcp snooping limit rate 10<br>ip dhcp snooping trust<br>!<br>interface fastethernet 0/24<br>switchport access vlan 303<br>ip dhcp snooping limit rate 10<br>ip dhcp snooping trust<br>!<br>&lt;省略部分&gt; | 配置信任端口 |

CORE1 运行配置：

| 配置 | 备注 |
|---|---|
| CORE1#show running-config<br>!<br>&lt;省略部分&gt;<br>interface fastethernet 0/22<br>switchport access vlan 301<br>ip dhcp snooping limit rate 10<br>ip dhcp snooping trust<br>!<br>interface fastethernet 0/23<br>switchport access vlan 302<br>ip dhcp snooping limit rate 10<br>ip dhcp snooping trust | |

| 配置 | 备注 |
|---|---|
| !<br>interface gigabitethernet0/1<br>switchport access vlan 300<br>ip dhcp snooping limit rate 10<br>ip dhcp snooping trust<br>!<br><省略部分> | 配置信任端口 |

CORE4 运行配置：

| 配置 | 备注 |
|---|---|
| CORE4#show running-config<br>!<br><省略部分><br>interface fastethernet 0/4<br>switchport access vlan 501<br>ip dhcp snooping limit rate 10<br>ip dhcp snooping trust<br>!<br>interface gigabitethernet0/1<br>switchport access vlan 300<br>ip dhcp snooping limit rate 10<br>ip dhcp snooping trust<br>!<br><省略部分> | 配置信任端口 |

⑤ 在相应网段进行测试。配置好设备和服务器，在终端用户验证是否能成功接收 DHCP 信息并隔离非法服务器的干扰，并测试网络。

⑥ 确认无误后，保存设备，并保存设备配置脚本文件。

# 项目 10　部署 DNS 系统

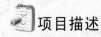

项目描述

域名系统（DNS）是最常用的系统之一，校园网用户浏览网页时基本都要使用 DNS 系统，如果都是用 ISP 的 DNS 服务器进行解析，出口解析流量将增大，给出口带来很大压力。同时，学校内部很多服务器都配置了域名，如果都是用 ISP 服务进行解析，很明显不合理。如何部署内部 DNS 服务器，同时解决外网解析的问题，将在本项目中讲解。

通过本项目，读者可以掌握以下技能：
- 能够分析 DNS 解析方法。
- 能够掌握申请 DNS 的方法及备案流程。
- 能够部署 DNS 转发。

知识准备

DNS 是 Domain Name System 的缩写，是一种由最多 255 个字符 128 层组织域组成的有层次结构的计算机和网络服务命名空间系统。当用户在应用程序中输入 DNS 名称时，DNS 服务可以将此名称解析为相关的 IP 地址信息。

1）根服务器

目前的域名解析采用的是"集中+分散解析"的方式，互联网管理委员会规定，全世界目前共有 13 个根服务器，负责全世界互联网域名解析工作。我们平时搭建的 DNS 服务器默认都指向这些根服务器，图 4-16 所示为目前全世界的 13 个根服务器及 IP 地址。

| 服务器完全合格的域名(FQDN) | IP 地址 |
|---|---|
| a.root-servers.net. | [198.41.0.4] |
| b.root-servers.net. | [128.9.0.107] |
| c.root-servers.net. | [192.33.4.12] |
| d.root-servers.net. | [128.8.10.90] |
| e.root-servers.net. | [192.203.230....] |
| f.root-servers.net. | [192.5.5.241] |
| g.root-servers.net. | [192.112.36.4] |
| h.root-servers.net. | [128.63.2.53] |
| i.root-servers.net. | [192.36.148.17] |
| j.root-servers.net. | [192.58.128.30] |
| k.root-servers.net. | [193.0.14.129] |
| l.root-servers.net. | [198.32.64.12] |
| m.root-servers.net. | [202.12.27.33] |

图 4-16  根服务器及 IP 地址

根服务器把以 com 结尾的域名的解析权委派给其他的 DNS 服务器，以后所有以 com 结尾的域名，根服务器就都不负责解析了，而由被委派的服务器负责解析。而且根服务器还把以 net、org、edu、gov 等结尾的域名都一一进行了委派，这些被委派的域名被称为顶级域名，每个顶级域名都有预设的用途，例如：com 域名用于商业公司，edu 域名用于教育机构，gov 域名用于政府机关等，这种顶级域名也被称为顶级机构域名。根服务器还针对不同国家进行了域名委派，例如：把所有以 CN 结尾的域名委派给中国互联网管理中心，以 JP 结尾的域名委派给日本互联网管理中心，CN、US 这些顶级域名被称为顶级地理域名，详见图 4-17。

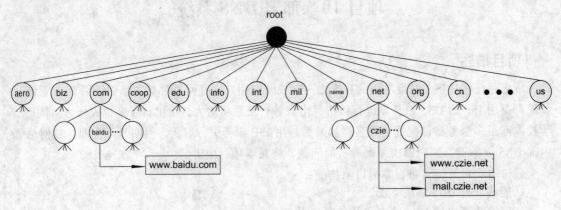

图 4-17  DNS 结构及委派图

2）DNS 解析

DNS 转发查询方式主要有迭代、递归两种，最常见的是迭代查询，下面结合图 4-18 以内网 PC0 查询 www.baidu.com 为例进行说明。

① 当用户 PC0 通过浏览器提出一个域名 www.baidu.com 解析要求时，会在系统的 DNS 缓存中查找是否有符合的记录信息，如果没有便会向自己系统中配置的本地（首选）DNS 服务器发起查询请求。

② 如果首选 DNS 服务器缓存内没有相应的记录信息，则会向全球离自己比较近的根服务器发起查询请求，这时根服务器会分析提交上来的域名，通过查询发现已经把以 com 结尾的所有域名解析的工作委派给了另一个下级的 DNS 服务器，则会把负责解析以 com 结尾的域名解析工作的这个 DNS 服务器的 IP 地址返回给提出查询请求的首选服务器。

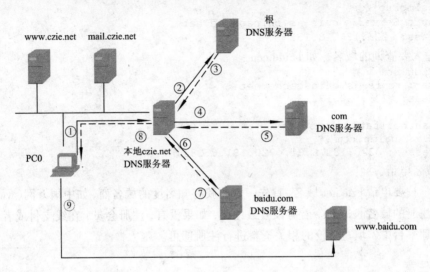

图 4-18　DNS 查询

③ 首选服务器会通过根服务器发来的 IP 找到下一级 DNS 服务器并提交 www.baidu.com 的查询请求，这时处在二级的服务器通过查询发现 www.baidu.com 服务器的 IP 记录。

④ PC0 获得域名和 IP 对应关系后，会把这个查询结果放入自己的缓存中，如果在缓存的有效期内有其他 DNS 用户再次请求这个域名，本地（首选）DNS 服务器就会利用自己缓存中的结果响应用户，而不用再去根服务器那里进行查询了。

3）DNS 记录说明

① A 记录（主机记录）：DNS 解析记录的基础，许多记录（像别名记录、NS 记录等）都依赖于它，并且 A 记录还可用做简易的负载均衡。

② NS 和 SOA 记录。
- NS 记录：描述域名服务器，指明区域中有几个 DNS 服务器。
- SOA 记录：起始授权记录，用于描述区域中的主 DNS 服务器。

③ Cname 记录：别名记录，用于给已存在的解析记录起别名用。

④ MX 记录：邮件交换器，用于邮件服务的记录。

⑤ SRV 记录：服务器位置记录（作用是指明区域内的 XXX 计算机提供 XXX 服务）。

## 项目实施

本项目通过部署内部 DNS 系统，来解决校园用户访问内外网时的地址解析问题。

1）域名系统分析

每个被委派的 DNS 服务器同样使用委派的方式向下发展，例如想申请使用 czie.net 域名，这时就要向负责 .net 域名的 DNS 服务器提出申请，只要 czie.net 还没有被其他单位或个人使用，而且申请者按时足额缴纳了费用，负责.net 域名的服务器就会把 czie.net 域名委派到该申请者。只要 DNS 服务器使用委派方式，域名空间就会逐步形成分布式解析架构。这种架构把域名解析权下放到自己的 DNS 服务器上，既有利于及时更新记录，同时对平衡流量压力也很有好处。

① 检查域名是否存在的方法：
```
C:>nslookup
Default Server:  cpis-js.chinacache.net
Address:  221.130.13.133
```
② 输入要查询的域名，如 baidu.com：
```
> baidu.com
Server:  cpis-js.chinacache.net
Address:  221.130.13.133

Non-authoritative answer:
Name:    baidu.com
Addresses:  220.181.6.81, 220.181.6.184
```
2）域名申请

如果学校要申请 czie.net 域名，首先一定要找有知名度的域名商，如中国万网（www.net.cn），查询所要注册的域名 czie.net，看有没有被注册，如果没有，注册会员，在线支付或者电汇费用，在会员管理平台中，填写学校的相关资料进行注册即可。

 注意

域名命名规则必须满足以下要求：

- 只能使用英文字母（a~z，不区分大小写）、数字（0~9），以及"-"（英文中的连字符），不能使用空格及特殊字符（如!、$、&、?等）。
- "-" 不能用做开头和结尾。
- 长度不能超过 63 个字符。

3）域名备案

如果校园网使用电信接入，根据工信部文件和各个地区要求，要求对 IP 和域名进行备案，参照文件为《工业和信息化部关于进一步落实网站备案信息真实性核验工作方案（试行）》（工信部电管【2010】64 号）。

4）DNS 部署 1：内网用户访问域名

校园网内终端用户首选 DNS 地址在 DHCP 部署中统一安排为：172.16.2.2。内部用户访问域名，会有两种类型：

- 访问内网服务器，如 www.czie.net、mail.czie.net。
- 访问外网服务器，如 www.baidu.com 等。

① 解决终端访问内网服务器问题，只需要在 172.16.2.2 服务器上配置两条 DNS 记录即可，详见表 4-4。

表 4-4　内部 DNS 记录

| 序 号 | 域 名 | 记 录 类 型 | 详 细 |
|---|---|---|---|
| 1 | www.czie.net | A | 172.40.0.1 |
| 2 | mail.czie.net | A | 172.20.1.1 |

② 解决终端访问外网域名的问题：由于172.16.2.2只能解析czie.net域内信息，终端用户访问外网域名，就需要DNS多级查询。

以baidu.com为例，.net和.com属于不同的顶级域名，因此需要查询根服务器，在172.16.2.2上配置根服务器记录，如表4-5所示。

表4-5 内部DNS记录

| 序 号 | 域 名 | 记 录 类 型 | 详 细 |
|---|---|---|---|
| 1 | Root | A | 202.12.27.33 |
| 2 | Com | NS | root |

③ root授权baidu.com由authority.baidu.com负责解析，需要在root上配置委派SOA、域名服务NS和主机A记录，如表4-6所示。

表4-6 DNS委派记录

| 序 号 | 域 名 | 记 录 类 型 | 详 细 |
|---|---|---|---|
| 1 | authority.baidu.com | A | 202.102.1.2 |
| 2 | baidu.com | NS | authority.baidu.com |
| 3 | Authority | SOA | authority |

④ 此时，baidu.com域名将由authority.baidu.com负责解析，需要在authority.baidu.com上配置委派SOA、主机A记录；如表4-7所示。

表4-7 baidu.com DNS记录

| 序 号 | 域 名 | 记 录 类 型 | 详 细 |
|---|---|---|---|
| 1 | authority.baidu.com | A | 202.102.1.2 |
| 2 | www.baidu.com | A | 220.181.6.81 |
| 3 | authority.baidu.com | SOA | authority |

⑤ 在PC0的浏览器中，输入"www.baidu.com"，成功访问，在172.20.2.2服务器上查看DNS缓存，如图4-19所示。在缓存有效期内，其他用户访问www.baidu.com将不再需要到根服务器去查找。

```
*(1) Name:baidu.com  Time Stamp:星期六 十月 16 13:57:39 2010
     Type: NS       server: authority.baidu.com
*(2) Name:www.baidu.com  Time Stamp:星期六 十月 16 13:57:39 2010
     Type: A        IP:          220.181.6.81
```

图4-19 本地DNS缓存记录

5）DNS部署2：外网用户访问czie.net服务器

校园网内网站服务器、电子邮件服务器需要提供外网访问功能，内部私有地址已经做了相应映射。在迭代解析方法上，请读者自行设计完成，要求在 PC12 上成功访问 czie.net 内网服务器，首选 DNS 为：202.102.1.2。

① 在 authority.baidu.com 上配置 root 服务器记录，填写表 4-8。

表 4-8　authority.baidu.com DNS 记录

| 序　号 | 域　名 | 记录类型 | 详　细 |
| --- | --- | --- | --- |
| | | | |
| | | | |
| | | | |

② root 授权 czie.net 由 authority.czie.net 负责解析，需要在 root 上配置委派 SOA、域名服务 NS 和主机 A 记录，填写表 4-9。

表 4-9　DNS 委派记录

| 序　号 | 域　名 | 记录类型 | 详　细 |
| --- | --- | --- | --- |
| | | | |
| | | | |
| | | | |

③ 此时，访问 czie.net 的解析工作将由 authority.czie.net 负责，需要在 authority.czie.net 上配置委派 SOA、主机 A 记录，填写表 4-10。

表 4-10　baidu.com DNS 记录

| 序　号 | 域　名 | 记录类型 | 详　细 |
| --- | --- | --- | --- |
| | | | |
| | | | |
| | | | |

6）DNS 生效

一般情况下，DNS 修改配置后不是马上生效，通常有一个时间差：
- 添加新的解析记录，生效时间是 10 分钟。
- 修改已经添加的解析记录，生效时间是 60 分钟。
- 修改为万网 DNS 服务器后首次进行域名解析，生效时间是 2 小时。
- 国内域名 DNS 修改，修改时间最长 6 小时。
- 国际域名 DNS 修改，修改时间最长 48 小时。

# 第 5 章 校园网出口设计

由于科研和对外通信的需要，校园网出口设计将成为校园网设计的重要环节之一。随着校园网用户量增加和基于校园网络的各种网络应用的展开，要求校园网络出口具有较高的带宽。许多学校采用了 CERNET 教育网和 TelCOM 电信（或联通、铁通等）的双出口方案，既可以保留教育网资源，同时可以增加带宽，提高访问 Internet 资源的速度。

内部私有服务器如果未进行公网地址映射，当其他校区或者外网用户访问这些服务器时，如何保证通信的成功和安全性，VPN 技术将是解决方案之一。

通过本章所有项目的实践，可以学会校园网出口的部署，通过配置出口策略路由、VPN、CBAC 防火墙，大大提高出口效率。

本章需要完成的项目有：

项目 11——部署 VPN；

项目 12——部署防火墙。

## 项目 11 部署 VPN

### 项目描述

图 5-1 所示为校园网 VPN 图，请进行部署以实现以下效果：

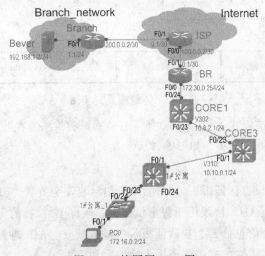

图 5-1 校园网 VPN 图

- 配置 IPSec VPN，实现校园网和分支服务器安全互联。
- 配置 SSL VPN，以便于外部用户安全接入校园网。

### 知识准备

VPN（Vistual Private Network）是一种日益受欢迎的技术，其原理是在公网上建立私密隧道进行通信。尤其是大型组织适用全球公网为基础进行通信，可以避免租用专线的高额费用，又能实现安全通信。

VPN 同时使用两种技术来保证隐私：IPSec（IP Security）、Tunneling（隧道技术）。

1）IPSec

IPSec 是一种由 IETF 设计，为数据包层在 IP 层提供安全的协议集。IPSec 没有定义任何特定封装或者认证方式的使用，然而它提供了一个框架和机制，关联了加密、身份验证选择、散列方法。

2）SA（安全关联）

IPSec 在两个主机之间进行逻辑连接使用信令协议：Security Assocation（SA），因为 IP 协议是面向无连接的，在保证安全前需要转变成面向连接的。SA 是一个源端和目标段的简单单向连接，如果双工连接需要两个 SA，每个方向一个 SA。SA 连接定义了三个要素：

① 一个 32 位的安全参数索引（SPI），如在帧中继和 ATM 等面向连接的协议中的标识符。
② 安全协议类型，IPSec 定义了两种可供选择的协议：AH 和 ESP。
③ 源 IP 地址。

3）两种模式

IPSec 工作在两种不同的模式：传输模式和隧道模式，模式定义了在 IP 报的哪一部分添加 IPSec 报头。

- Transport Mode（传输模式）：在该模式下，IPSec 报头将添加在 IP 报头之后，如图 5-2 所示。

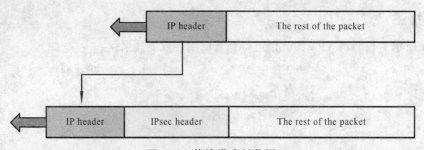

图 5-2 传输模式封装图

- Tunnel Mode（隧道模式）：在该模式下，IPSec 报头将添加在源 IP 报头之前，在 IPSec 报头前再添加新的 IP 报头，如图 5-3 所示。

4）两个安全协议

IPSec 定义了两个协议：认证报头（AH）协议和封装安全负载（ESP）协议。

Authentication Head（AH）协议用于源主机认证，确保 IP 封装的负载的完整性，使用哈希功能和一个对称密钥计算信息摘要，在认证报头中插入摘要。AH 通常放在基于传输模式的合适位置。图 5-4 显示了在传输模式中认证报头的内容和位置。

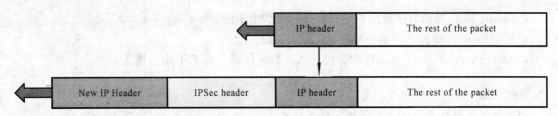

图 5-3　隧道模式封装图

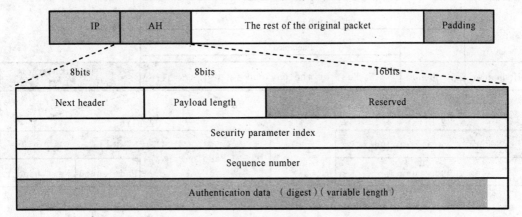

图 5-4　AH 图

AH 主要字段如下。
- Next header：使用 8 位空间定义原 IP 包负载协议类型（如 TCP、UDP、ICMP、OSPF 等），其功能等同于封装前 IP 报头中的协议类型，换言之，将原来的协议字段值复制到现在位置。
- Payload length：该 8 位空间的名称存在误导，不是定义了负载的长度，而是定义 AH 的长度。
- SPI 专有 32 位值：与目的 IP 地址和安全协议（AH）相结合，唯一识别数据报的安全关联（Security Association）。
- Sequence number：32 位，包含无变化的增长计数器值，该值是强制性的，即使接收端不为特定 SA 提供 Anti-Replay 服务，它仍然存在。
- Authentication data：一个可变长字段，包括在 ESP 数据包上计算的减去 Authentication data 的完整校验值（ICV）。

 注意

AH 协议提供信息认证和完整性，对整个 IP 包进行处理，没有加密特性格式。

Encapsulating Security Payload（ESP）协议增加了加密特性，ESP 的认证包添加到数据包末端，计算更加简单。图 5-5 显示了在传输模式中认证报头的内容和位置。

ESP 主要字段如下。
- SPI 专有 32 位值：功能类似于 AH 相应字段。
- Sequence number：32 位，功能类似于 AH 相应字段。
- Padding：被保护的数据，加密算法需要一个 initialization vector（IV），注意 IV 需要认证。DES 使用前 8 个字节作为 IV，3DES、AES 也使用 8 字节的 IV。

- Pad length：根据加密算法不同，补足的字结也不同。
- Next header：8位，功能类似于 AH 相应字段。
- Authentication data：在 AH 中，包含认证数据计算，在 ESP 中，则无。

> **注意**
> ESP 协议提供信息认证、完整性、加密特性。

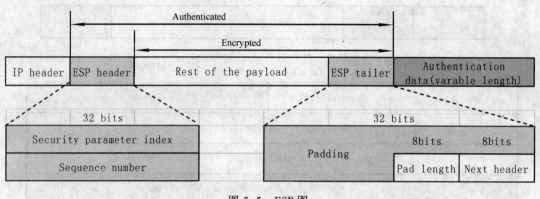

图 5-5　ESP 图

ESP、AH 对比：

① AH 没有 ESP 的加密特性。

② AH 的 authtication 是针对整个数据包的，包括 IP 头部分，因为 IP 头部分包含很多变量，比如 type of service（TOS）、flags、fragment offset、TTL 以及 header checksum，所以这些值在进行 authtication 前要全部清零，否则会导致丢包。

5）Tunneling

VPN 规定，数据封装必须在另一个数据包中，如图 5-6 所示。

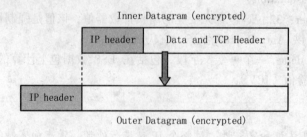

图 5-6　Tunneling 图

### 项目实施

1）IPSec VPN 准备

保障校园网私有地址用户与分校区服务器安全通信。

① 按照拓扑图，配置各个设备的 IP 地址。

② 校园网总部使用 OSPF 方式进行配置。

③ 在 PC0（172.16.0.2）上 ping WEB 服务器（192.168.1.2），看路由是否连通，因为 ISP 缺乏路由，所以无法连通，如图 5-7 所示。

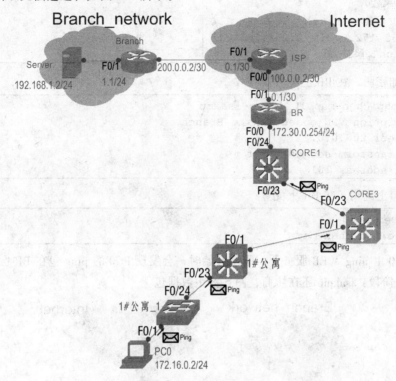

图 5-7　PC0 ping WEB 服务器

2）IPSec VPN 部署

① 定义策略：

| crypto isakmp policy 10 | 双方的优先级可以不同，系统会自动协商<br>数值越低，优先级越高 |
|---|---|
| lifetime 3600 | 双方的密钥生存周期若不一致，则以短的一方为准 |
| encryption 3des | 加密协议：3des |
| hash md5 | 认证方式：md5 |
| authentication pre-share | 共享密钥 |

② 设置密钥（key）：

| crypto isakmp key chen-pwd address 200.0.0.2 | 配置密钥为 chen-pwd，对端 IP：200.0.0.2（在 Branch 上配置） |
|---|---|

③ 设置传输集：

| crypto ipsec transform-set chen-set esp-3des esp-3des esp-md5-hmac | 可以设多个 transform，必须一致，名称（chenset）可以不同 |
|---|---|

④ 设置感兴趣的机密数据流：

| acc 101 permit ip 172.0.0.0 0.255.255.255 192.168.1.0 0.0.0.255（在 BR 上配置）<br>acc 101 permit ip 192.168.1.0 0.0.0.255 172.0.0.0 0.255.255.255（在 Branch 上配置） | 两端 ACL 必须对称；否则加密过程完成不了 |
| --- | --- |

⑤ 创建加密图。在 BR 上配置：

```
crypto map chen-map 10 ipsec-isakmp
  description VPN conection to Branch
  set peer 200.0.0.2
  set transform-set chen-set
  match address 101
```

⑥ 在接口上应用。在 BR 上配置：

```
Interface fa0/0
  Crypto map chen-map
```

⑦ 在 PC0 上 ping WEB 服务器，刚刚开始时，会发现 PC0 的 ping 包在 BR 会丢弃，因为 VPN 的第一个阶段：isakmp 还在协商，图 5-8 中数据包。

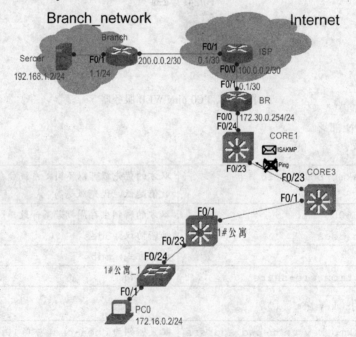

图 5-8　PC0 Ping WEB 服务器丢包

⑧ 分析 ISAKMP 数据包如，图 5-9 所示。

⑨ 因为在 BR 转发出去时，源地址和目标地址都已经改变，重新进行封装，详见图 5-10、图 5-11、图 5-12。

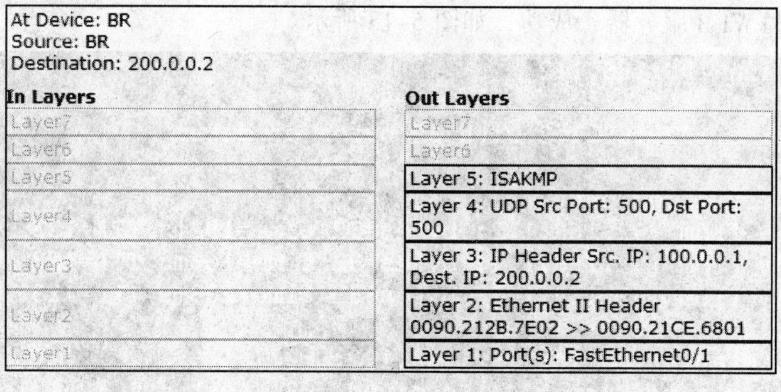

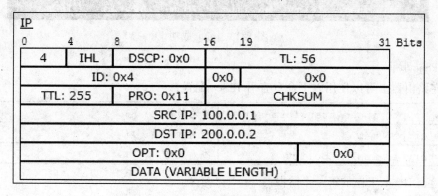

图 5-9 分析包

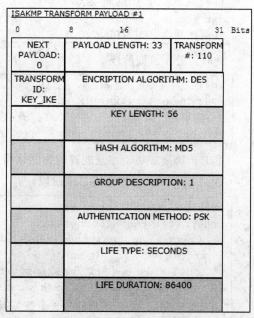

图 5-10 IP 封装图

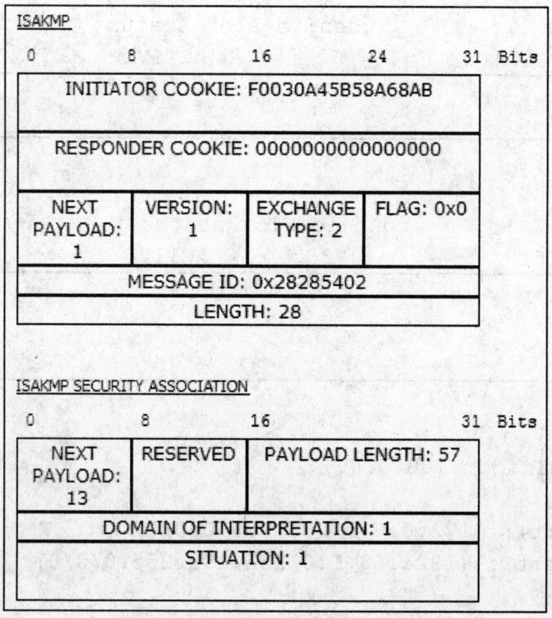

图 5-11 ISAKMP 封装图 1        图 5-12 ISAKMP 封装图 2

⑩ 再次 ping WEB 服务器，成功，如图 5-13 所示。

```
PC>ping 192.168.1.2

Pinging 192.168.1.2 with 32 bytes of data:

Request timed out.
Reply from 192.168.1.2: bytes=32 time=192ms TTL=123
Reply from 192.168.1.2: bytes=32 time=166ms TTL=123
Reply from 192.168.1.2: bytes=32 time=203ms TTL=123

Ping statistics for 192.168.1.2:
    Packets: Sent = 4, Received = 3, Lost = 1 (25% loss),
Approximate round trip times in milli-seconds:
    Minimum = 166ms, Maximum = 203ms, Average = 187ms
```

图 5-13　ping 成功

3）IPSec 验证

① 在路由器上可以使用以下命令查看 IPSec 配置：

| show crypto ipsec transform-set | 查看传输集 |
| --- | --- |
| show crypto map | 查看加密图 |
| show crypto isakmp policy | 查看策略 |

② 检查 IPSec 工作的第一个阶段。

在 BR 上运行 debug crypto isakmp sa 命令：

```
BR#show crypto isakmp sa
IPv4 Crypto ISAKMP SA
dst             src             state           conn-id slot status
200.0.0.2       100.0.0.1       MM_NO_STATE     0       0    ACTIVE
```

此时，处于协商阶段。

```
BR#show crypto isakmp sa
IPv4 Crypto ISAKMP SA
dst             src             state           conn-id slot status
200.0.0.2       100.0.0.1       QM_IDLE         1092    0    ACTIVE
```

此时，协商成功，注意配置的为默认路由。

③ 检查 IPSec 工作的第二个阶段：

```
BR#show crypto ipsec sa
interface: FastEthernet0/1
  Crypto map tag: chen-map, local addr 100.0.0.1
  protected vrf: (none)
  local  ident (addr/mask/prot/port): (172.0.0.0/255.0.0.0/0/0)
  remote ident (addr/mask/prot/port): (192.168.1.0/255.255.255.0/0/0)
  current_peer 200.0.0.2 port 500
   PERMIT,  flags={origin_is_acl, }
  #pkts encaps: 1, #pkts encrypt: 1, #pkts digest: 0
  #pkts decaps: 1, #pkts decrypt: 1, #pkts verify: 0
```

```
  #pkts compressed: 0,  #pkts decompressed: 0
  #pkts not compressed: 0,  #pkts compr. failed: 0
  #pkts not decompressed: 0,  #pkts decompress failed: 0
  #send errors 1,  #recv errors 0
    local crypto endpt.: 100.0.0.1, remote crypto endpt.:200.0.0.2
    path mtu 1500, ip mtu 1500, ip mtu idb FastEthernet0/1
    current outbound spi: 0x008954A2(9000098)
    inbound esp sas:
     spi: 0x1E343F60(506740576)
       transform: esp-3des esp-md5-hmac ,
       in use settings ={Tunnel, }
       conn id: 2008, flow_id: FPGA:1, crypto map: chen-map
       sa timing: remaining key lifetime (k/sec): (4525504/3474)
       IV size: 16 bytes
       replay detection support: N
       Status: ACTIVE
    inbound ah sas:
    inbound pcp sas:
    outbound esp sas:
     spi: 0x008954A2(9000098)
       transform: esp-3des esp-md5-hmac ,
       in use settings ={Tunnel, }
       conn id: 2009, flow_id: FPGA:1, crypto map: chen-map
       sa timing: remaining key lifetime (k/sec): (4525504/3474)
       IV size: 16 bytes
       replay detection support: N
       Status: ACTIVE
    outbound ah sas:
    outbound pcp sas:
```

此时，IPSec 验证成功。

4）SSL VPN 部署

目前，SSL VPN 主要的应用场合是将园区外的用户安全接入园区网络，如学校领导在校外会见客人时需要访问校园网的信息，这时 SSL VPN 就是最好的解决方案。

CISCO 的 SSL 可以使用 2811 路由器或者 ASA 防火墙来实现，本项目通过路由器说明，因为 Packet 不能完成配置，重点讲解 SSL VPN 的使用方法及步骤。

如图 5-14 所示，保障在校园内移动用户安全接入校园网。

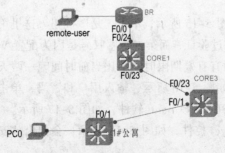

图 5-14　SSL VPN 部署图

（1）配置 VPN 服务器（BR）：

| | |
|---|---|
| BR(config)# aaa new-model<br>BR(config)# aaa authentication login default local | 为防止控制台超时而造成无法进入 Exec |
| BR(config))# aaa authentication login webvpn local<br>BR(config)# ip local pool ssl-add 11.1.1.1 11.1.1.200<br>BR(config)# username useBR password 123 | 定义 WebVPN 本地认证用户名、密码 |
| BR(config))# webvpn gateway vpngateway | 定义 WebVPN 在哪个接口上进行监听,此时 IOS 会自动产生自签名证书 |
| BR (config-webvpn-gateway)# ip address 192.168.10.10 port 443<br>BR (config-webvpn-gateway)# inservice | 启用 WebVPN gateway 配置 |
| BR (config)# webvpn context webcontext | 定义 WebVPN 的相关配置,相当于 ASA 的 tunnel-group,在这里可以定义 |
| BR (config-webvpn-context)# gateway vpngateway | 将 context 和 gateway 相关联 |
| BR (config-webvpn-context)# aaa authentication list webvpn<br>BR (config-webvpn-context)# inservice | 启用 WebVPN context 配置 |
| BR(config-webvpn-context)# policy group sslvpn-policy | 进入 sslvpn 策略组 |
| BR(config-webvpn-group)# functions svc-enabled<br>BR(config-webvpn-group)# svc address-pool ssl-add | 分配 svc 使用的地址池 |
| BR(config-webvpn-group)# svc split include 192.168.20.0 255.255.255.0 | 定义隧道分离的目标地址,如果不配置,则默认为 0.0.0.0 |
| BR(config-webvpn-group)#exit<br>BR(config-webvpn-context)# default-group-policy sslvpn-policy | 当配置了多个 policy group 后,默认使用的策略组 |

（2）VPN 客户端验证（remote-uer）

① 在浏览器中输入"https://192.168.10.10/"访问 WebVPN,这时会弹出提示信息,单击"确定"按钮。

② 需要安装证书,如图 5-15 所示,单击"是"按钮。这里第一个感叹号是因为这个证书是路由器自签发的,没有经过验证,而第二个感叹号是因为配置 WebVPN 时应该注意颁发后的有效期,往往颁发证书时的有有效期限时间会比当前时间晚一两天。

③ 这时会弹出网页,如图 5-16 所示,输入用户和密码,单击 Login 按钮。

④ 这时会提示安装 SSL VPN Client 软件,如图 5-17 所示。

⑤ 如果允许安装 Active 控件,如图 5-18 所示,会弹出"安全警告"对话框,单击"安装"按钮。

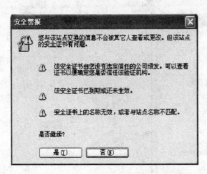
图 5-15 "安全警报"对话框

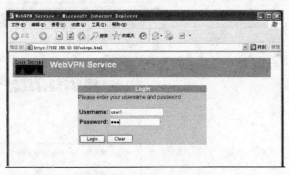
图 5-16 输入用户名和密码

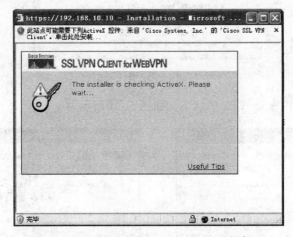
图 5-17 提示安装 SSL VPN Client 软件

图 5-18 "安全警告"对话框

⑥ SSL VPN Client 安装进程如图 5-19 所示。
⑦ 出现"证书"对话框时,单击"安装证书"按钮,如图 5-20 所示。

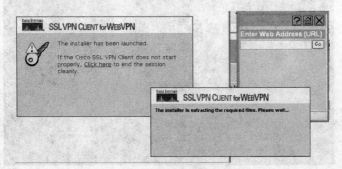
图 5-19 安装过程

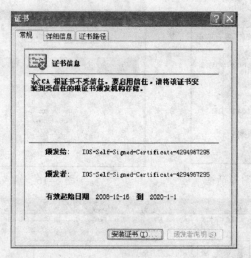

图 5-20 "证书"对话框

⑧ 在客户端上可以查看 VPN 的状态，如图 5-21 所示。

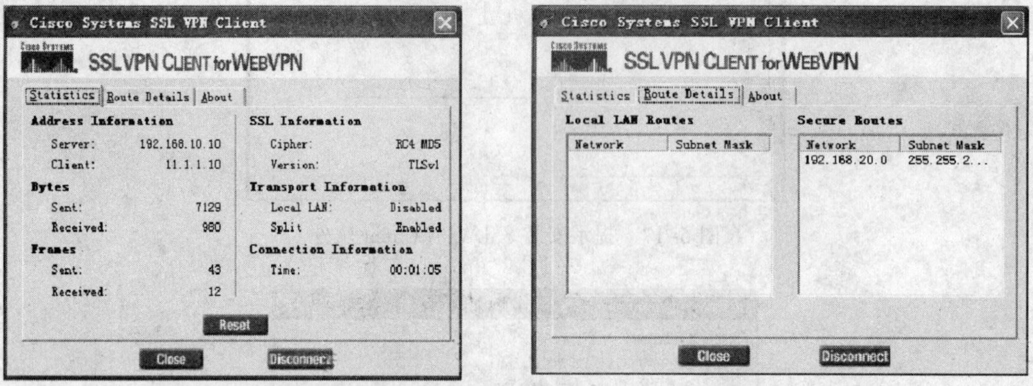

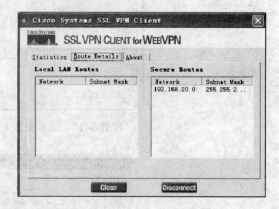

图 5-21 在 SSL 客户端查看 VPN 状态

⑨ 使用 ipconfig 命令可以查看到获得的地址，如图 5-22 所示。

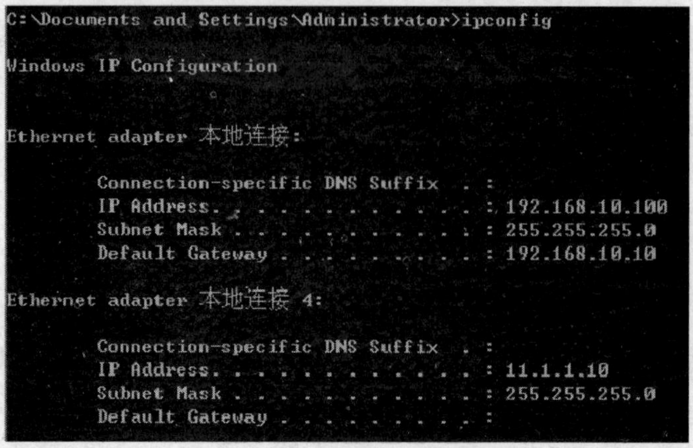

图 5-22 常看获得的地址

## 工程化操作

① 初始化设备，保证三层路由联通。

对 BR 初始化如下：

```
##################    br-init-cfg    ##################
configure terminal
hostname Branch
enable secret en_pwd
interface fastethernet 0/0
 no shutdown
 ip address 200.0.0.2 255.255.255.252
 exit
interface fastethernet 0/1
 no shutdown
 ip address 192.168.1.1 255.255.255.0
 exit
line vty 0 4
 login
 password vty_pwd
 end
```

② 创建配置脚本。新建文本文件，将各个设备配置命令复制，粘贴到其中，保存文件名为：设备名_ VPN_CFG.txt。

创建脚本文件为 BR_VPN_CFG.txt，内容如下：

```
##################    br-vpn-cfg    ##################
configure terminal
interface fastethernet 0/0
 no shutdown
 ip add 172.30.0.254 255.255.255.0
 exit
interface fastethernet0/1
 no shutdown
 ip add 100.0.0.1 255.255.255.252
 exit
ip route 172.0.0.0 255.0.0.0 172.30.0.1
ip route 10.0.0.0 255.0.0.0 172.30.0.1
ip route 0.0.0.0 0.0.0.0 100.0.0.2
crypto isakmp policy 110
 authentication pre-share
 encryption des
 group 1
 hash md5
 lifetime 86400
 exit
crypto isakmp key chen-pwd address 200.0.0.2
crypto ipsec transform-set chen-set esp-3des esp-md5-hmac
access-list 101 permit ip 172.0.0.0 0.255.255.255 192.168.1.0 0.0.0.255
crypto map chen-map 10 ipsec-isakmp
 set transform-set chen-set
```

```
 set peer 200.0.0.2
 match address 101
 set pfs group1
 exit
interface fastethernet 0/1
 crypto map chen-map
 exit
```

创建脚本文件为 Branch_VPN_CFG.txt，内容如下：

```
################## branch-vpn-cfg ##################
configure terminal
interface fastethernet 0/0
 no shutdown
 ip address 192.168.1.1 255.255.255.0
 exit
interface fastethernet0/1
 no shutdown
 ip add 100.0.0.2 255.255.255.252
 exit
ip route 0.0.0.0 0.0.0.0 200.0.0.1
crypto isakmp policy 110
 authentication pre-share
 encryption des
 group 1
 hash md5
 lifetime 86400
 exit
crypto isakmp key chen-pwd address 200.0.0.1
crypto ipsec transform-set chen-set esp-3des esp-md5-hmac
access-list 101 permit ip 192.168.1.0 0.0.0.255 172.0.0.0 0.255.255.255
crypto map chen-map 10 ipsec-isakmp
 set transform-set chen-set
 set peer 200.0.0.1
 match address 101
 set pfs group1
 exit
interface fastethernet 0/1
 crypto map chen-map
 exit
```

③ 执行配置脚本：使用 CONSOLE 口或者远程连接登录设备，打开超级终端或者 SecureCRT 程序，进入特权模式，复制设备中的命令脚本，在超级终端或者 SecureCRT 程序中，右击选择"粘贴"，执行脚本。

④ 验证配置：在各设备的特权模式下，使用 show running-config 命令检查当前运行配置。

⑤ 在相应网段进行测试，检查 VPN 是否部署成功，确认无误后，保存设备，并保存设备配置脚本文件。

# 项目12  部署防火墙

## 项目描述

校园网的内 WEB、DNS 服务提供外网用户访问，出于安全的考虑，杜绝其他来自外网的访问流量，同时内网用户要能自由访问外网所有资源，使用 ACL 无法实现，可以配置 CBAC 来完成。

## 知识准备

CBAC（context-based access control）是 Cisco IOS 防火墙特性集的一个高级防火墙模块，即基于上下文的访问控制，它不同于 ACL（访问控制列表），并不能用来过滤每一种 TCP/IP 协议，但它对于运行 TCP、UDP 应用或某些多媒体应用（如 Microsoft 的 NetShow 或 Real Audio）的网络来说是一个较好的安全解决方案。除此之外，CBAC 在流量过滤、流量检查、入侵检测等方面表现卓越。在大多数情况下，我们只需在单个接口的一个方向上配置 CBAC，即可实现只允许属于现有会话的数据流进入内部网络。可以说，ACL 与 CBAC 是互补的，它们的组合可实现网络安全的最大化。

Cisco 路由器的 access-list 只能检查网络层或者传输层的数据包，而 CBAC 能够智能过滤基于应用层的 TCP 和 UDP 的 session，能够在 firewall access-list 打开一个临时的通道给起源于内部网络向外的连接，同时检查内外两个方向的 sessions。CBAC 可提供如下服务。

① 状态包过滤：对企业内部网络、企业和合作伙伴互联以及企业连接 Internet 提供完备的安全性和强制政策。

② Dos 检测和抵御：CBAC 通过检查数据报头、丢弃可疑数据包来预防和保护路由器受到攻击。

③ 实时报警和跟踪：可配置基于应用层的连接，跟踪经过防火墙的数据包，提供详细过程信息并报告可疑行为。

## 项目实施

① 在出口路由器 BR 上配置 ACL。

外部访问 ACL：

```
BR(config)#ip access-list extended ext_acl
BR(config-ext-nacl)#permit tcp any host 172.16.2.2 eq 53
BR(config-ext-nacl)#permit udp any host 172.16.2.2 eq 53
BR(config-ext-nacl)#permit tcp any host 172.40.0.1 eq 80
BR(config-ext-nacl)#deny ip any any
```

DMZ 访问 ACL：

```
BR(config)# ip access-list extended dmz_acl
BR(config-ext-nacl)# permit icmp any any
BR(config-ext-nacl)# permit tcp any any
```

内部访问 ACL：

```
BR(config)#ip access-list extended int_acl
BR(config-ext-nacl)#deny ip any any
```

② 定义 CBAC 检查规则：

```
BR(config)# ip inspect audit-trail
BR(config)#ip inspect audit-trail
BR(config)#ip inspect name dmz_cbac icmp
BR(config)#ip inspect name dmz_cbac tcp
BR(config)#ip inspect name int_cbac icmp
BR(config)#ip inspect name int_cbac tcp
```

③ 在接口上应用 ACL 与检查：

```
BR(config)# interface FastEthernet1/0
BR(config-if)# ip access-group ext_acl in
BR(config-if)# exit
BR(config)# interface FastEthernet1/1
BR(config-if)# ip access-group dmz_acl in
BR(config-if)# ip inspect dmz_cbac in
BR(config-if)# exit
BR(config)# interface FastEthernet0/0
BR(config-if)# ip access-group int_acl in
BR(config-if)# ip inspect int_cbac in
BR(config-if)# exit
BR(config)# interface FastEthernet0/1
BR(config-if)# ip access-group int_acl in
BR(config-if)# ip inspect int_cbac in
BR(config-if)# exit
```

④ 在外网 www.baidu.com 上访问 172.40.0.1 WEB 服务器，如图 5-23 所示。

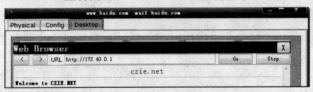

图 5-23　成功访问 WEB 服务器

⑤ 测试到 BR 路由器的连通性：ping58.0.0.2，无法访问，如图 5-24 所示。

图 5-24　ping BR 路由器失败

⑥ 在 PC0 上访问 www.baidu.com，成功！

在 BR 上出现提示：

```
%FW-6-SESS_AUDIT_TRAIL_START: Start tcp session: initiator (172.16.0.1:1028)
-- responder (220.181.6.81:80)
%FW-6-SESS_AUDIT_TRAIL_STOP: Stop tcp session: initiator (172.16.0.1:1028)
-- responder (220.181.6.81:80)
```

使用 show ip inspect sessions detail 查看连接状态表的统计信息，包括所有会话：

```
BR#show ip inspect sessions detail
Established Sessions
 Session 150210136 (172.16.0.1:1029)=>(220.181.6.81:23) tcp SIS_OPEN
  Created 00:00:05, Last heard 00:00:04
  Bytes sent (initiator:responder) [360:360]
  In  SID 220.181.6.81[80:80]=>172.16.0.1[1029:1029] on ACL ext_acl
  (3 matches)
```

⑦ 在 PC0 上 Telnet 58.0.0.1（见图 5-25），成功！

图 5-25　成功 Telnet 到 58.0.0.1

在 BR 上出现提示：

```
%FW-6-SESS_AUDIT_TRAIL_STOP: Stop tcp session: initiator (172.16.0.1:1029)
-- responder (220.181.6.81:80)
%FW-6-SESS_AUDIT_TRAIL_START: Start tcp session: initiator (172.16.0.1:1030)
-- responder (58.0.0.1:23)
```

使用 show ip inspect sessions detail 查看连接状态表的统计信息，包括所有会话：

```
BR#show ip inspect sessions detail
Established Sessions
 Session 150210136 (172.16.0.1:1031)=>(58.0.0.1:23) tcp SIS_OPEN
  Created 00:00:02, Last heard 00:00:02
  Bytes sent (initiator:responder) [360:360]
  In  SID 58.0.0.1[23:23]=>172.16.0.1[1031:1031] on ACL ext_acl (2 matches)
```

通过以上几种流量的分析，会发现 CBAC 维持具有连接信息的会话状态表，当状态表中的一个条目表明此分组属于某个被允许的会话时，会在防火墙中制造一个动态的通路，供返回流量使用。

## 工程化操作

① 初始化设备，保证三层路由连通。

② 在出口 BR 上部署 CBAC，创建配置脚本。新建文本文件，将各个设备配置命令复制、粘贴到其中，保存文件名为：设备名_cbac_CFG.txt。

创建脚本文件为 br_cbac_CFG.txt，内容如下：

```
################## br-cbac -cfg ##################
configure terminal
ip access-list extended ext_acl
 permit tcp any host 172.16.2.2 eq 53
 permit udp any host 172.16.2.2 eq 53
 permit tcp any host 172.40.0.1 eq 80
 deny ip any any
 exit
ip access-list extended dmz_acl
 permit icmp any any
 permit tcp any any
 exit
ip access-list extended int_acl
 permit ip any any
 exit
ip inspect audit-trail
ip inspect name dmz_cbac icmp
ip inspect name dmz_cbac tcp
ip inspect name int_cbac icmp
ip inspect name int_cbac tcp
interface FastEthernet1/0
 ip access-group ext_acl in
 exit
interface FastEthernet1/1
 ip access-group dmz_acl in
 ip inspect dmz_cbac in
 exit
interface FastEthernet0/0
 ip access-group int_acl in
 ip inspect int_cbac in
 exit
interface FastEthernet0/1
 ip access-group int_acl in
 ip inspect int_cbac in
 exit
```

③ 执行配置脚本：使用 CONSOLE 口或者远程连接登录设备，打开超级终端或者 SecureCRT 程序，进入特权模式，复制设备中的命令脚本，在超级终端或者 SecureCRT 程序中，右击选择"粘贴"，执行脚本。

④ 验证配置：在各设备的特权模式下，使用 show running-config 命令检查当前运行配置，无误后保存设备配置。

# 第6章 校园网运行维护

校园网用户数量庞大，用户的不当操作、病毒等都可能造成网络不稳定，甚至出现故障，所以网络的日常运行维护尤为重要，采用哪些措施能有效提高校园网的效率是校园网建设的一个重要环节。

本章需要完成的项目有：

项目13——使用管理工具；

项目14——AAA部署。

## 项目13　使用管理工具

### 项目描述

对于校园网来讲，设备数量众多，在应用策略做调整或者进行故障修复时，管理员每次都到各个楼宇的设备间直接使用CONSOLE口配置设备，不太现实，通常会使用远程方式进行管理。一般情况下，在设备安装进机柜前，通常会把常规的配置（包含远程配置）设置好，在物理层联通的情况下，就可以远程对校园网的所有设备进行配置。

在校园网的运维过程中，经常会出现策略的调整，在测试策略过程中，通常需要对原来的策略（设备配置）进行备份，适当的时候可能需要恢复。通常会部署TFTP或者FTP服务器。

在较小的网络中，网络设备比较少的情况我们常常是用Telnet、SSH或者Web等方法对网络设备进行维护和监控。但随着网络规模的逐渐增大，网络设备的数量增加，网络管理员很难及时监控所有设备的状态，发现并修复故障；网络设备也可能来自不同的厂商，这将使网络管理变得更加复杂。在这种情况下，可以使用SNMP协议来统一管理。

图1-3为校园网拓扑图，为便于校园网运维和管理，提出以下任务：

① 远程管理校园网中的交换、路由设备。

② 对校园网策略进行备份、恢复灵活。

③ 通过SNMP、镜像端口管理校园网。

### 知识准备

1）Telnet

Telnet是一种客户端/服务器协议，规定了创建和终止虚拟终端（VTY）会话的规则。该协议还规定了启动Telnet会话的命令语法和顺序，以及在会话过程中可以使用的控制命令。每个

Telnet 命令都包含至少两个字节，第一个字结是特殊字符，称为命令字符（IAC），正如其名称所示，IAC 规定其下一个字结必须是命令而非文本。

Telnet 命令包括：

① Are You There（AYT）——允许用户请求终端屏幕上显示的资源，以表明 VTY 会话处于活动状态。

② Erase Line（EL）——从当前行中删除所有文本。

③ Interrupt Process（IP）——暂停、中断、放弃或者终止与虚拟终端相连的进程。例如，如果用户通过 VTY 在 Telnet 服务器上打开一个程序，他/她也可以发出一条 IP 命令来终止程序。

尽管 Telnet 协议支持用户身份验证，但是它不支持加密（encrypted）数据的传输。所有在 Telnet 会话期间交换的数据都将以纯文本格式在网络内传输，这样的话，传输的数据可能会被中途截获并被读取。

2）SSH

安全外壳（SSH）协议是一种更安全的远程设备访问方法。此协议提供与 Telnet 相似的远程登录结构，但使用更安全的网络服务。

SSH 提供比 Telnet 更严格的口令身份验证，并在传输会话数据时采用加密手段。SSH 会话会将客户端与 IOS 设备之间的所有通信加密，使得用户 ID、口令和管理会话的详细信息保持私密。作为一种最佳实践，只要可能，就应该始终用 SSH 替代 Telnet。

大多数新版 IOS 包含 SSH 服务器。在某些设备中，此服务默认启用。还有些设备需要启用 SSH 服务器才能启用此服务。IOS 设备还配备了 SSH 客户端，该客户端可用于与其他设备建立 SSH 会话。同理，还可以使用运行 SSH 客户端的远程计算机来启动安全 CLI 会话。并非所有的计算机操作系统都默认提供 SSH 客户端软件。

3）FTP /TFTP

FTP（文件传输协议）能提供主机到主机的文件传输，不同于其他的 C/S 应用的是：在主机之间使用两个连接，一个传输数据，使用 TCP 20 号端口，另一个控制信息，使用 TCP 21 号端口。数据和控制独立连接，能保证 FTP 的高效性，如图 6-1 所示。

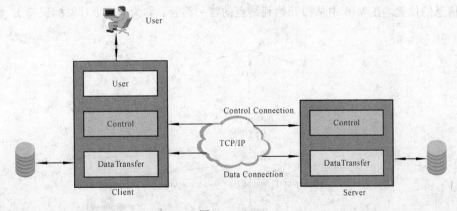

图 6-1　FTP

控制连接一直持续在整个 FTP 会话过程中，数据连接针对每个文件传输连接、断开，在一个控制连接期间，数据连接可能会多次连接、断开。

控制连接中，服务器开放 21 号端口，等待客户端使用随机端口连接，促使 FTP 延迟很低；数据连接中，客户端使用随机端口发送给服务端，服务端收到信息后使用 20 号端口连接、传输数据。

TFTP（简单文件传输协议）比 FTP 简单，用于无盘工作站、路由器启动或读取配置等，仅仅用来快速传输文件。TFTP 使用 UDP 69 号端口。

4）SNMP

SNMP（简单网络管理协议）使用 TCP/IP 协议族对网络上的设备进行管理，提供一组基本的操作，用来监视和管理网络。

在 PC 上安装一个网络管理软件（像 CISCO 的 CISCO network，Windows Server 自带的网络管理软件等）作为网络管理站，网络设备就是被管理对象，网络管理软件与网络设备之间的通信用到的就是 SNMP 协议。通过 SNMP 协议可以获得网络设备名、网络配置情况、物理接口状态等，通过 SNMP 协议可以用网络管理软件对网络设备进行配置，当网络发生问题时，网络设备通过 SNMP 及时反馈到网络管理站，网络管理员就能第一时间知道网络出现了问题。

网络管理软件通过 SNMP 实现监控和配置网络设备的原理是：用网络管理软件获得、配置、监控网络设备，其实就对网络设备的数据库的获取、配置、监控。这种数据库就是 MIB（管理信息数据库）。在网络管理软件中也有和网络设备相对的 MIB，可能不同厂商都对标准的 MIB 做了一定"私有化"，只要将这些私有 MIB 导入网络管理软件，我们就能配置、管理网络设备。当网络管理站（也就是装了网络管理软件的 PC）对自己的 MIB（与被操作网络设备 MIB 一致）进行操作（如 Get——获取网络设备的信息，Set——修改网络设备配置）时，产生的应用数据通过 SNMP 协议传递。数据按 IOS 的七层自上而下封装，数据到达被管理的网络设备解封，最终查看到是 SNMP 数据包，根据数据包里的命令网络设备对自己的 MIB 进行操作（如将自己的设备信息发送给网络管理站等）。

MIB 是一种数据库，它存放了网络设备的各种信息，如系统命令信息、配置信息、接口状态信息、路由信息等。MIB 是以树状结构进行存储的，树的结点表示被管理的对象（如网络设备的接口信息等）。也就是说，要对 MIB 进行操作，就必须在这个树状的数据存结构中找到所要管理信息的位置。在 MIB 中从根开始到结点的唯一路径，称为 OID（对象标识符），如图 6-2 所示。

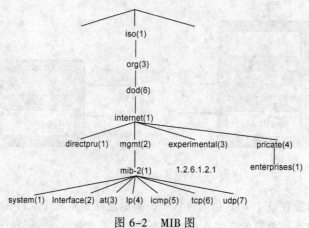

图 6-2　MIB 图

# 项目实施

1）Telnet 部署

在前面第 2 章 VLAN 部署项目中，我们已经体验了通过管理 VLAN 管理设备的方法；在管理交换机路由器时，通常使用三层（接口 IP）地址。试想一下：

如果远程管理设备在其他网段，而因路由故障，又无法直接 ping 通接口地址，使用接口 IP Telnet 应该说已经失去意义。在前面的项目中我们对设备管理 VLAN 了进行规划，如表 6-1 所示。

表 6-1 VLAN 规划表

| 楼 宇 | 部 门 | 设 备 | VLAN ID | VLAN NAME | IP | 子 网 掩 码 |
|---|---|---|---|---|---|---|
| 1#公寓 | 1#公寓 | DS_1_GY | V99 | manage | 10.0.0.10 | 255.255.255.0 |
| | | AS_1_GY_1 | V99 | manage | 10.0.0.11 | 255.255.255.0 |
| | | AS_1_GY_2 | V99 | manage | 10.0.0.12 | 255.255.255.0 |
| 2#公寓 | 2#公寓 | DS_2_GY | V99 | manage | 10.0.0.20 | 255.255.255.0 |
| | | AS_2_GY_1 | V99 | manage | 10.0.0.21 | 255.255.255.0 |
| | | AS_2_GY_2 | V99 | manage | 10.0.0.22 | 255.255.255.0 |
| 3#公寓 | 3#公寓 | DS_3_GY | V99 | manage | 10.0.0.30 | 255.255.255.0 |
| | | AS_3_GY_1 | V99 | manage | 10.0.0.31 | 255.255.255.0 |
| | | AS_3_GY_2 | V99 | manage | 10.0.0.32 | 255.255.255.0 |
| 4#公寓 | 4#公寓 | DS_4_GY | V99 | manage | 10.0.0.40 | 255.255.255.0 |
| | | AS_4_GY_1 | V99 | manage | 10.0.0.41 | 255.255.255.0 |
| | | AS_4_GY_2 | V99 | manage | 10.0.0.42 | 255.255.255.0 |
| 教学楼 | 教学楼 | DS_JX | V99 | manage | 10.0.0.50 | 255.255.255.0 |
| | | AS_JX_1 | V99 | manage | 10.0.0.51 | 255.255.255.0 |
| | | AS_JX_2 | V99 | manage | 10.0.0.52 | 255.255.255.0 |
| 实训楼 | 实训楼 | DS_SX | V99 | manage | 10.0.0.60 | 255.255.255.0 |
| | | AS_SX_1 | V99 | manage | 10.0.0.61 | 255.255.255.0 |
| | | AS_SX_2 | V99 | manage | 10.0.0.62 | 255.255.255.0 |
| 图书馆 | 图书馆 | DS_TSG | V99 | manage | 10.0.0.70 | 255.255.255.0 |
| | | AS_TSG_1 | V99 | manage | 10.0.0.71 | 255.255.255.0 |
| | | AS_TSG_2 | V99 | manage | 10.0.0.72 | 255.255.255.0 |
| 综合楼 | 综合楼 | DS_ZH | V99 | manage | 10.0.0.80 | 255.255.255.0 |
| | | AS_ZH_1 | V99 | manage | 10.0.0.81 | 255.255.255.0 |
| | | AS_ZH_2 | V99 | manage | 10.0.0.82 | 255.255.255.0 |
| 主楼 | 网络中心 | CORE1 | V99 | manage | 10.0.0.91 | 255.255.255.0 |
| | | CORE2 | V99 | manage | 10.0.0.92 | 255.255.255.0 |
| | | CORE3 | V99 | manage | 10.0.0.93 | 255.255.255.0 |
| | | CORE4 | V99 | manage | 10.0.0.94 | 255.255.255.0 |

在图 1-3 的基础上，对校园网进行简化，详见图 6-3。下面我们将使用 Telnet 工具来远程管理网络。

① 配置网管员工作站地址为：10.0.0.100/24，与其他设备的管理地址在相同网段。

② 配置各个设备，以 DS_4_GY 设备为例：

```
DS_4_GY(config)#vlan 99
DS_4_GY(config-vlan)#name manage_vlan
```

创建 VLAN99，并命名

```
DS_4_GY(config)#interface range fastethernet 0/1,fastethernet 0/24
DS_4_GY(config-range-if)#switchport trunk encapsulation dot1q
DS_4_GY(config-range-if)#switchport mode trunk
```

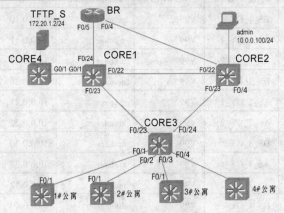

图 6-3 校园网简化图

进行 TRUNK 封装，3560 需要手工配置封装协议

```
DS_4_GY(config-range-if)#switchport trunk native vlan 99
```

设置 native vlan 为 99

```
DS_4_GY(config)#interface vlan 99
DS_4_GY(config-if)#no shutdown
DS_4_GY(config-if)#ip address 10.0.0.40 255.255.255.0
```

配置 VLAN99 地址

```
DS_4_GY(config)#enable secret cisco
```

配置特权密码

```
DS_4_GY(config)#line vty 0 4
DS_4_GY(config-line)#login
DS_4_GY(config-line)#pass cisco
```

配置 VTY 线路，同时支持 5 条；配置 Telnet 登录；配置 Telnet 登录密码

③ 在 PC0（172.16.0.2）上 ping Web 服务器（192.168.1.2），看路由是否连通，因为 ISP 缺乏路由，所以无法连通。

④ 在 Admin 上进行 Telnet 测试，如图 6-4 所示。

图 6-4  Telnet 测试

> **注意**
> Telnet 使用 TCP 23 端口进行通信，在部署 ACL 时，注意不能将 23 号端口封掉！

2）SSH 部署

通过 Telnet 进行远程管理的便捷性显而易见，但 Telnet 最大的问题是：数据以明文方式进行传输，安全性较低。通常使用另一种方式进行替换：SSH。

SSH 为 Secure Shell 的缩写，由 IETF 的网络工作小组（Network Working Group）所制定。SSH 是建立在应用层和传输层基础上的安全协议。SSH 是目前较可靠，专为远程登录会话和其他网络服务提供安全性的协议。利用 SSH 协议可以有效防止远程管理过程中的信息泄露问题。

① 配置各个设备以 DS_4_GY 设备为例：

```
DS_4_GY(config)#ip domain-name czie
```

配置域名为 czie

```
DS_4_GY(config)#crypto key generate rsa
```

为路由器的 SSH 加密会话产生加密密钥

```
D S_4_GY(config)#username xzchen password cisco
```

配置账户（xzchen）和密码（cisco）

```
DS_4_GY(config)#line vty 0 4
DS_4_GY(config-line)#transport input ssh
```

出于安全角度考虑，仅允许使用 SSH 方式登录

```
DS_4_GY(config-line)#login local
```

使用本地数据库进行验证

② 设置密钥（key）。在 Admin 上登录测试，方法如下：

```
c:\ssh -l [用户名] [IP 地址]
```

用户名：刚刚创建的 xzchen；IP 地址：10.0.0.40。

在 cmd 下，输入：

```
ssh -l xzchen 10.0.0.40
```

提示输入密码，即为 xzchen 的密码：cisco。

详细效果请见图 6-5。

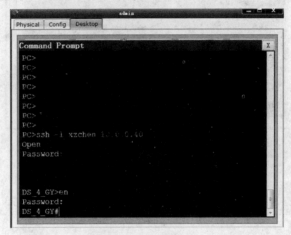

图 6-5　SSH 测试

 注意

SSH 默认使用 TCP 22 端口进行通信，在部署 ACL 时，注意不能将 22 号端口封掉！

3）备份/恢复

在校园网的运维过程中，经常会出现策略的调整，在测试策略过程中，通常需要对原来的策略（设备配置）进行备份，适当的时候可能需要恢复。通常会部署 TFTP 或者 FTP 服务器。

① 在数据中心部署 TFTP 服务器，便于备份和恢复设备配置。

② 核心层 CORE1 因策略调整，要求将现有配置备份到 TFTP 服务器上。

③ 配置路由，使得 CORE1 能 ping 通 TFTP 服务器。详细内容见第 3 章 OSPF 相关项目。

④ 在 TFTP 服务器上 ping "CORE1" 的 IP 地址（10.8.0.1、10.8.1.1 或者 10.8.2.1），而不能再使用：10.0.0.91，请分析原因。

⑤ 在 CORE1 上将 startup-config 备份到 TFTP：

```
core1# copy startup-config tftp:
Address or name of remote host []? 172.20.1.2
```

输入服务器地址

```
Destination filename [core1-confg]? 20111109-core1-config
```

备份文件名为 20111109-core1-config

```
Writing startup-config...!!
[OK - 1770 bytes]
```

备份成功，文件大小为 1770B

⑥ 在 TFTP 服务器可以查看，如图 6-6 所示。

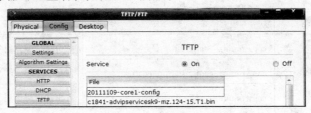

图 6-6　查看 TFTP 配置

⑦ 当下次要恢复时，可以执行下列命令：

```
core1#copy tftp: startup-config
Address or name of remote host []? 172.20.1.2
```

输入服务器地址

```
Source filename []? 20111109-core1-config
```

输入备份文件名

```
Destination filename [startup-config]?
```

恢复到启动配置文件

```
Accessing tftp://172.20.1.2/20111109-core1-config...
Loading 20111109-core1-config from 172.20.1.2: !
[OK - 1770 bytes]
1770 bytes copied in 0.048 secs (36875 bytes/sec)
```

从 TFTP 下载，成功

⑧ 使用 FTP 备份，方法类似于 TFTP，FTP 的功能比 TFTP 更强大，尤其针对用户权限问题。在 FTP 服务器上创建账户：xzchen，密码：cisco，如图 6-7 所示。

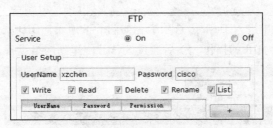

图 6-7　查看 FTP 配置

⑨ 将 running-config 备份到 FTP 服务器：

```
core1#copy running-config ftp: xzchen cisco
```

 注意

该命令在 Packet tracer 5.3 无法实现，如需配置请使用真实设备。

存储器之间传输通常会有 4 个存储区域，在空间运行的情况下，它们之间是可以相互传输（备份/恢复），可以使用通用命令进行：

Copy 空间 1　空间 2

其中，空间 1 为源位置，空间 2 为目标位置，详见表 6-2。

表 6-2 存储空间介绍

| 空间 | 存储内容 | 命令 | 备注 |
|---|---|---|---|
| RAM | running-config | copy runing-config tftp: | 备份运行配置到 TFTP |
| | | copy runing-config startup-config | 保存 |
| | | copy running-config flash | 备份运行配置到 FLASH |
| FLASH | IOS | copy flash: running-config | 运行 FLASH 中配置 |
| | | copy flash: startup-config | 恢复 FLASH 中配置到启动配置 |
| | | copy flash: tftp: | 保存 FLASH 到 TFTP |
| NVRAM | startup-config | copy startup-config running-config | 运行启动配置 |
| | | copy startup-config flash: | 备份启动配置到 FLASH |
| | | copy startup-config tftp: | 备份启动配置到 TFTP |
| TFTP/FTP | 备份 | copy tftp: startup-config | 恢复配置到启动配置 |
| | | copy tftp: running-config | 恢复配置到运行配置 |
| | | copy tftp: flash | 恢复 TFTP 到 FLASH |

4）SNMP 部署

若要使用 SNMP，需要先在设备上启动，本项目以 CORE2 为例进行介绍。

① 开启 SNMP，创建用户如表 6-3 所示。

表 6-3 创建用户

| snmp-server community xzchen RO | 创建只读用户：xzchen |
|---|---|
| snmp-server community cxz RW | 创建读/写用户：cxz |

② 在 admin 对话框中进入 Destop 选项卡，如图 6-8 所示。OID 就是 MIB 的信息结点精确路径，Operations 就是对 MIB 的操作，然后通过 SNMP 传递到网络设备上。

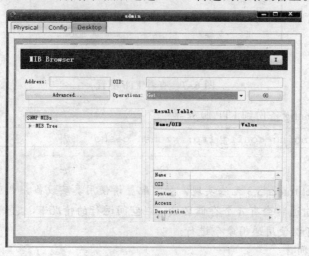

图 6-8 Desktop 选项卡

③ 单击 Advanced 按钮，填入需要管理网络设备的 IP 地址，以 Community 值，用于网络设备与网络管理设备的身份认证，如图 6-9 所示。Address 是设备的管理 IP，read community 是 xzchen，write community 是 cxz。

图 6-9 Advanced 对话框

④ 在 MIB 数据库中精确查找我们需要的信息结点路径，再通过 Get 操作获取网络设备的信息。如图 6-10 所示，在 MIB 中找到相应的结点路径（OID：.1.3.6.1.2.1.1.1.0 或者 iso.org.dod.interface.mgmt.mib-2.system.sysDescr），执行 Get 操作，获得 CORE2 的系统信息描述信息，包括交换机的型号和 IOS 型号等。

> **注意**
> Access 项显示 read-only，说明该信息是只读信息，不能通过 SNMP 来配置网络设备，若为 write-read 就可配置。

图 6-10 获取网络设备信息

⑤ 通过 SNMP 配置网络的 hostname。同样，先找到 MIB 中 sysname 的位置（.1.3.6.1.2.1.1.5.0），然后获取网络设备的 hostname，如图 6-11 所示。

图 6-11　获取 hstname

⑥ 看到值跟 CORE2 上命名一致。下面修改网络设备的 hostname，改成 3560，如图 6-12 所示。

⑦ 将 Operations 改为 Set，获取 OID 位置，数据类型是字符串，Value 就是修改的值 3560，然后单击 OK 按钮，再单击 GO 按钮。在回到 CORE2 上，在命令行中一次按【Enter】键就看到效果了，发现主机名被修改了，如图 6-13 所示。

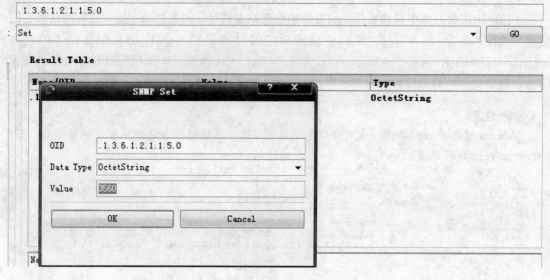

图 6-12　修改 hostname

总结：

基于 SNMP 的网管软件可以方便地对网络设备进行监控和管理。采用 SNMP 来对网络进行控制和监控的系统，也称为 NMS（Network Management System）。

常用的运行在 NMS 上的网管平台有 HP OpenView、CiscoView、CiscoWorks 2000、Star View。

```
core2>
core2>
3560>
3560>
3560>
```

图 6-13　修改效果

### 工程化操作

① 保存 DS_4_GY 当前配置为备份文件，采用 copy running-config flash 命令。目标文件为：DS_4_GY_cfg_bk。

② 对 DS_4_GY 进行 Telent 配置。

创建脚本文件为 DS_4_GY_TELNET_CFG.txt，内容如下：

```
##################    ds-4-gy-telnetl-cfg    ##################
configure terminal
vlan 99
 exit
interface range fastethernet 0/1,fastethernet 0/24
 switchport mode trunk
 switchport trunk native vlan 99
 exit
interface vlan 99
 no shutdown
 ip address 10.0.0.40 255.255.255.0
 exit
enable secret cisco
line vty 0 4
 login
 password cisco
 end
```

③ 执行配置脚本。使用 CONSOLE 口或者远程连接登录设备，打开超级终端或者 SecureCRT 程序，进入特权模式，复制设备中的命令脚本，在超级终端或者 SecureCRT 程序中，右击并选择"粘贴"命令，执行脚本。

④ 验证配置。在各设备的特权模式下，使用 show running-config 命令检查当前运行配置；在 Admin 上 Telnet 到远程设备 DS_4_GY。其他设备配置方法同上。

⑤ 对 DS_4_GY 进行 SSH 配置。

创建脚本文件为 DS_4_GY_SSH_CFG.txt，内容如下：

```
##################    ds-4-gy-ssh-cfg    ##################
configure terminal
ip domain-name czie
crypto key generate rsa
line vty 0 4
 transport input ssh
 login local
 end
```

⑥ 执行配置脚本：使用 CONSOLE 口或者远程连接登录设备，打开超级终端或者 SecureCRT 程序，进入特权模式，复制设备中的命令脚本，在超级终端或者 SecureCRT 程序中，右击选择"粘贴"，执行脚本。

⑦ 验证配置。在各设备的特权模式下，使用 show running-config 命令检查当前运行配置；在 Admin 上 SSH 登录到远程设备 DS_4_GY。其他设备配置方法同上。

⑧ 对 CORE1 进行 SSH 配置。新建文本文档 RSPAN_CORE1_CFG.txt，添加命令形成配置

脚本（脚本中去除模式）。

创建脚本文件为 CORE1_RSPAN_CFG.txt，内容如下：

```
##################    core1-rspan-cfg    ##################
configure terminal
vlan 100
 remote-span
 exit
monitor session 1 source interface FastEthernet 0/23
monitor session 2 destination remote vlan 100
reflector-port FastEthernet 0/21
interface FastEthernet 0/22
 switchport mode trunk
 exit
```

⑨ 对 CORE2 进行 RSPAN 配置。新建文本文档 RSPN_CORE2_CFG.txt，添加命令形成配置脚本（脚本中去除模式）。

创建脚本文件为 CORE2_RSPAN_CFG.txt，内容如下：

```
##################    core2-rspan-cfg    ##################
configure terminal
monitor session 1 source int fa0/22 - 24 both
monitor session 1 destination int fa0/1
monitor session 2 source remote vlan 100
monitor session 2 destination interface FastEthernet 0/1
```

⑩ 执行配置脚本：使用 CONSOLE 口或者远程连接登录设备，打开超级终端或者 SecureCRT 程序，进入特权模式，复制设备中的命令脚本，在超级终端或者 SecureCRT 程序中，右击选择"粘贴"，执行脚本。检查 RSPAN，测试是否成功！

# 项目 14　AAA 部署

## 项目描述

我们可以以 Telnet、SSH、Web 或者 CONSOLE 口等不同方式登录交换机、路由器获得普通或者特权模式的访问权限。在网络设备众多、对网络安全要求较高的校园网中，仅仅使用交换机本地数据库的账号、密码认证功能已经满足不了用户的要求，管理员往往希望能够采用集中、多重的认证方式，这样也能在提高安全度的前提下减轻网络管理负担。RADIUS 服务就是其中之一。

所有思科交换机、路由器进行统一的登录认证。具体希望实现以下功能：

① 无论是通过 CONSOLE 口还是通过 Telnet，登录用户必须首先通过 AAA 认证，只有数据库中的授权账户才能访问交换机。

② 认证的用户统一管理，分配设备查看、修改配置的权限。

图 6-14 所示为校园网 AAA 图。

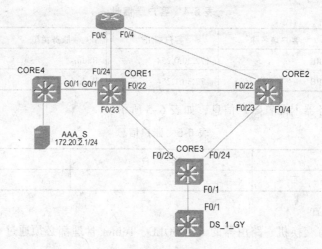

图 6-14　AAA 图

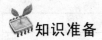 **知识准备**

AAA 即 authentication——认证，authorization——授权，accounting——记账，RADIUS（remote authentication dial in user service）由 RFC2865、RFC2866 定义，是目前应用最广泛的 AAA 协议，最初是由 Livingston 公司提出的，原来的目的是为拨号用户进行认证和计费。后来经过多次改进，形成了一项通用的认证计费协议。

RADIUS 是一种 C/S 结构的协议，它的客户端最初就是 NAS（net access server）服务器，现在任何运行 RADIUS 客户端软件的计算机都可以成为 RADIUS 的客户端。RADIUS 协议认证机制灵活，可以采用 PAP、CHAP 或者 UNIX 登录认证等多种方式。RADIUS 是一种可扩展的协议，它进行的全部工作都是基于 Attribute-Length-Value 的向量进行的。 RADIUS 的基本工作原理：用户接入 NAS，NAS 向 RADIUS 服务器使用 Access-Require 数据包提交用户信息，包括用户名、密码等相关信息，其中用户密码是经过 MD5 加密的，双方使用共享密钥，这个密钥不经过网络传播；RADIUS 服务器对用户名和密码的合法性进行检验，必要时可以提出一个 Challenge，要求进一步对用户认证，也可以对 NAS 进行类似的认证；如果合法，给 NAS 返回 Access-Accept 数据包，允许用户进行下一步工作，否则返回 Access-Reject 数据包，拒绝用户访问；如果允许访问，NAS 向 RADIUS 服务器提出计费请求 Account-Require，RADIUS 服务器响应 Account-Accept，对用户的计费开始，同时用户可以进行自己的相关操作。

RADIUS 工作详细过程，读者可参考相关教程或者 RFC，由于篇幅问题，本书不再细述。

**项目实施**

1）部署 AAA 服务器

① 在 AAA 服务器上配置客户端信息，如表 6-4 所示。

表6-4 客户端信息

| 序号 | 客户端名称 | 客户端IP | 服务类型 | 密钥 |
|---|---|---|---|---|
| 1 | BR | 172.30.0.254 | Radius | 123 |
| 2 | BR | 172.30.1.254 | Radius | 123 |

② 在AAA服务器上配置账户信息，如表6-5所示。

表6-5 账户信息

| 序号 | 账户 | 密码 |
|---|---|---|
| 1 | Xzchen | cisco |

2）认证设备配置

本项目要求所有交换机、路由器的CONSOLE、Telnet管理都必须通过AAA服务器认证方可登录、管理。

① 启用AAA认证：

```
aaa new-model
```

全局下，启用AAA功能

② 设置认证模式：

```
aaa authentication login default group radius none
```

配置默认认证方式为radius

```
aaa authentication login telnet_lines group radius
```

配置telnet_lines认证方式为radius

③ 配置服务器地址：

```
radius-server host 172.20.2.1 auth-port 1645 key xzchen
```

配置服务器地址、端口和密钥

④ 配置CONSOLE口验证：

```
line con 0
login authentication default
```

使用默认方式验证

⑤ 配置VTY验证：

```
line vty 0 4
login authentication telnet_lines
```

VTY启用telnet_lines认证

通常园区网会部署认证方式，尤其在用户访问特殊网络（如互联网）时，会进行计费等，常见的系统有Dr.com。

### 工程化操作

① 保存BR当前配置为备份文件，使用copy running-config flash命令。目标文件为：

BR_cfg_bk1。

② 配置脚本 BR AAA。

创建脚本文件为 BR_AAA_CFG.txt，内容如下：

```
##################   br-aaa-cfg   ##################
configure terminal
aaa new-model
aaa authentication login default group radius none
aaa authentication login telnet_lines group radius
radius-server host 172.20.2.1 auth-port 1645 key xzchen
line con 0
 login authentication default
 exit
line vty 0 4
 login authentication telnet_lines
 exit
```

③ 执行配置脚本。使用 CONSOLE 口或者远程连接登录设备，打开超级终端或者 SecureCRT 程序，进入特权模式，复制设备中的命令脚本，在超级终端或者 SecureCRT 程序中，右击并选择"粘贴"命令，执行脚本。

④ 验证配置。在各设备的特权模式下，使用 show running-config 命令检查当前运行配置。

⑤ 在相应网段进行测试，验证实际效果，确认无误后，保存设备，并保存设备配置脚本文件。

# 第 7 章 IPv6 部署

IPv4 地址枯竭这一问题在 20 世纪 80 年代就被注意到了。尽管一种替代 IPv4 的技术已经研发出来，但是互联网工程师们在当时还是找到了几种能够保持互联网基于 IPv4 继续增长的技术。

这些技术包括可将私人网络地址隐藏在一个单一的公共 IPv4 地址后面的网络地址转换（NAT）技术和可更为有效地分配和路由 IP 地址段的无类别域间路由（CIDR）。CIDR 和 NAT 技术的应用使得互联网工程协会推迟了将互联网基础设施由 IPv4 升级至 IPv6 的进程。

目前运营商和企业正在逐渐向 IPv6 升级。地区性互联网注册机构（RIR）在 2010 年 4 月份表示，目前尚未分发出去的 IPv4 地址数量已经低于 8%。

IPv4 地址空间将在 2011 年被用尽。在校园网部署 IPv6 可以更好地适应网络发展趋势。本章需要完成的项目有：

项目 15——IPv6 实验网。

## 项目 15 IPv6 实验网

### 项目描述

图 7-1 所示为校园网 IPv6 拓扑图，请按照要求规划：

① 配置 IPv6 静态路由。
② 配置 IPv6 和 IPv4 转换（NAT-PT）。

### 知识准备

1）IPv6 过渡方案

互联网工程协会已经部署了多个过渡方案，这些过渡方案将帮助网络使用者逐渐由 IPv4 向 IPv6 迁移。其中一个方案被称为"双堆栈"机制，其允

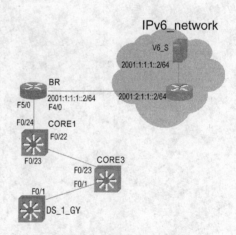

图 7-1 校园 IPv6 拓扑图

许网络使用者在他们的网络中同时使用 IPv4 和 IPv6。另一个方案被称为运营商级网络地址转换（CGN），其允许多个用客户共享一个单一的公共 IPv4 地址。第三个方案被称为"6rd"。该方案已经被法国 ISP 运营商 Free 所使用。法国 ISP 运营商 Free 通过在基于 IPv4 骨干网络的传输中封装 IPv6 数据包快速地向家庭用户部署了 IPv6。最终用户可以在多种隧道技术中做出选择。这

些隧道技术包括"6to4 技术"、"Teredo 技术"和"ISATAP 技术"。虽然这些隧道技术均可以在 IPv4 网络中传输 IPv6 数据包，但是这些技术也存在安全方面的隐患。

2）IPv6 地址规划

目前全球可用的 IPv4 网址剩余量不足 10%，随着网络的普及，手机、笔记本、服务器、路由器等设备，都在消耗 IP 地址，3G 网络和互联网的推广和应用更进一步加大了 IP 地址的消耗。中国的 IPv4 资源分配只占全球的 4.5%，第 25 次中国互联网络发展状况统计报告显示，中国网民达到 3.84 亿，成长速度为 28.9%，IP 地址增幅和数量落后于网民增幅和数量，所以中国面临的形势更为严峻，面对紧缺的 IPv4 资源，网络管理者应立即采取措施，推进 IPv6 网络部署规划。

为简化 IPv6 地址申请流程，促进 IPv6 的申请和应用，APNIC 在 2010 年 2 月推出了 IPv6 地址申请的快速通道计划，根据这一计划，凡是持有 IPv4 地址且尚未持有 IPv6 地址的会员，可以免费申请一块 IPv6 地址（默认大小：/32）。就目前的情况而言，全球分配的 IPv6 地址块 80%的大小是/32，也就是说，就目前可以预见的情况，/32 的 IPv6 地址块基本能满足大部分企业的需求，因此快速通道计划采用/32 作为默认大小，不做进一步审核，大大简化了申请流程。

IPv6 地址空间巨大，如果按 IPv4 的地址使用方法，显然一块/32 地址（2$^{96}$ 个 IP 地址）就能满足全世界的需求，然而，正是因为 IPv6 地址空间巨大，可以允许层次化、结构化的 IP 地址规划和分配，便于路由汇聚，从而可以大大缩小路由表，并且简化网络管理、配置、变更和扩展的工作量。对于大型互联网服务商，采用层次化、结构化的 IP 地址规划和分配，/32 的地址量就不一定能满足网络部署的需求。

值得注意的是，如果互联网服务商已经先申请了一块/32 的 IPv6，如果在后期的网络规划中发现需要更多的 IPv6 地址，再申请时就需要经过严格的审核流程，特别是要考查原先的 IPv6 地址的使用率。在 IPv6 网络规划和部署初期，由于用户量少，使用率很难满足后续申请的需求，从而对网络地址规划造成困难，即使在若干年后使用率达到要求，申请后续的 IPv6 地址，也有可能跟原来的/32 是不连续的地址段，造成路由汇聚的困难。

（1）层次化设计

大型互联网服务商的 IPv6 网络采用层次化、结构化的设计，可以把网络分为多个区域（地区），每个区域有多个区域骨干结点，每个骨干结点汇聚多个接入网（站）点，通过接入网点连接终端网点（企业或个人用户）提供服务。一般来说，是一个 4 层级的网络模型，而小型互联网服务商则采用 1~2 层级的网络模型。

（2）IPv6 地址结构

为了便于路由汇聚和简化网络管理、配置、变更和扩展的工作量，IPv6 采用层次化、结构化的地址规划。一般来说，按 RFC3177 的原则，把 128 位的 IPv6 地址分为一个 64 位的网络标识符和一个 64 位的结点（主机）标识符，结点（主机）标识符通常根据物理地址自动生成，叫做 EUI-64（或者 64-位扩展唯一标识）；64 位的网络标识符又进一步分为全球网络标识符和本地子网标识符，通常全球网络标识符占用 48 位，本地子网标识符占用 16 位，如图 7-2 所示。

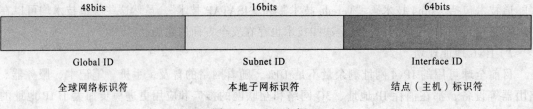

图 7-2 IPv6 空间 1

按照图 7-2 的原则，终端网点分配一段/48 的 IPv6 地址，最终连接主机/PDA 设备的子网分配一段/64 的 IPv6 地址，假设互联网服务商自己分配到的 IPv6 地址段是/32，则该服务商有 48-32=16 位用做自己服务网络的网点标识，总共可以允许 65 536 个网点标识，如图 7-3 所示。

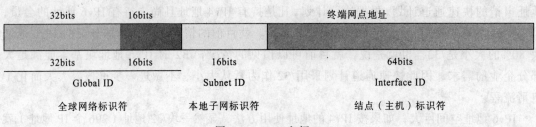

图 7-3 IPv6 空间 2

65 536 个网点数量上是很大，但如果再考虑以上所述的层次化设计，服务商有 16 位用做自己服务网络的网点标识，如果要结构化地把这 16 位用于上面所述的 4 层（区域、骨干结点、接入网点、终端网点）网络模型，平均每个层级可以占用 4 位（注：本文描述的是一个通用的模型，因此采用平均分配的方式，实际设计应该根据每个层级预期的结点数量决定所占用的位数），这样每层的结点数就只可以有 16 个，如图 7-4 所示。

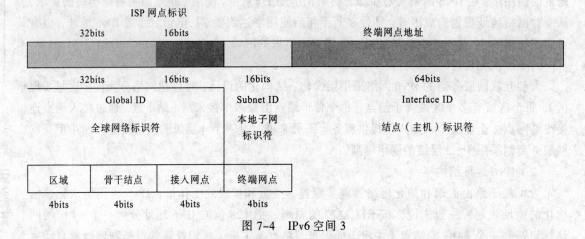

图 7-4 IPv6 空间 3

可见，对大型的全国性的互联网服务商，如果网络层次较多，一个/32 的地址就相对不足。

当然，这主要还是根据网络的层次和终端网点的地址分配策略来定，例如，如果上述终端网点分配从/48 改为/56，互联网服务商的网点标识就有 56-32=24 位，每个层级有 6 位可用，允许 64 个结点，这样/32 就相对够用。同样，如果网络是 2~3 层结构的话，/32 也相对够用。

（3）IPv6 地址分配策略

一般来说，使用以下分配策略分配 IPv6 地址：

① 1/64 分配给最终连接主机、PDA 等终端设备的子网。

② 1/56 分配给家庭用户，小型办公/家庭办公（SOHO）、远程办公分支机构。

③ 1/48 分配给企业客户。

如图 7-5 所示，根据以上分配策略，一个/32 的 IPv6 地址块，总共可以支持 65 536 个/48（216 个）终端用户，或 16777216 个/56（224 个）终端用户。

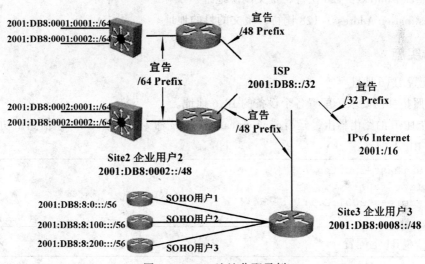

图 7-5 IPv6 地址分配示例

3）IPv6 封装格式

IPv6 封装格式如图 7-6 所示。

| Version | Traffic Class | Flow Label | |
|---|---|---|---|
| Payload Length | | Next Header | Hop Limit |
| Source Address | | | |
| Destination Address | | | |

图 7-6 IPv6 封装格式

各个字段的意义如下。

① Version：4 比特，值为 6 表示 IPv6 报文。

② Traffic Class：8 比特，类似于 IPv4 中的 TOS 字段。

③ Flow Label：20 比特，IPv6 中新增。可用来标记特定流的报文，以便在网络层区分不同的报文。转发路径上的路由器可以根据 Flow Label 来区分流并进行处理。由于 Flow Fabel 在 IPv6 报文头中携带，转发路由器可以不必根据报文内容来识别不同的流，目的结点也同样可以根据

Flow Label 识别流，同时由于 Flow Label 在报文头中，因此使用 IPSec 后仍然可以根据 Flow Label 进行 QoS 处理。

④ Payload Length：16 比特，以字节为单位的 IPv6 载荷长度，也就是 IPv6 报文基本头以后部分的长度（包括所有扩展头部分）。

⑤ Next Header：8 比特，用来标识当前头（基本头或扩展头）后下一个头的类型。此字段定义的类型与 IPv4 中的协议字段值相同。IPv6 定义的扩展头由基本头或扩展头中的扩展头域链接成一条链。这一机制下处理扩展头更高效，转发路由器只处理必须处理的选项头，提高了转发效率。

⑥ Hop Limit：8 比特，和 IPv4 中的 TTL 字段类似。每个转发此报文的结点把此字段值减 1，如果此字段值减到 0 则丢弃。

⑦ Source Address：128 比特，报文的源地址。

⑧ Destination Address：128 比特，报文的目的地址。

## 项目实施

（1）配置 IPv6 地址

① 按照拓扑图要求，配置各个设备的 IPv6 地址。

在配置 IPv6 的路由器上，运行 ipv6 unicast-routing 命令，开启 IPv6 单播路由功能。

② 在 BR 上配置：

```
ipv6 unicast-routing
interface fa0/1
ipv6 address 2001:2:1:1::1/64
exit
```

③ 在 V6_R1 上配置：

```
hostname V6_R1
ipv6 unicast-routing
interface fa0/1
  ipv6 address 2001:2:1:1::2/64
  exit
interface fa0/0
  ipv6 address 2001:1:1:1::1/64
  exit
```

（2）配置 IPv6 静态路由

① 在 BR 上配置 IPv6 静态路由，方法同 IPv4 静态路由：

```
ipv6 route 2001:1:1:1::/64 2001:2:1:1::2
```

② 在 V6_R1 上配置 IPv6 静态路由：

```
ipv6 route ::/0 2001:2:1:1::1
```

注意，配置的为默认路由。

③ 在 BR 上使用 show ipv6 router 命令验证 IPv6 路由：

```
BR#sh ipv6 route
IPv6 Routing Table - 3 entries
Codes:C - Connected, L - Local, S - Static, R - RIP, B - BGP
```

```
          U - Per-user Static route, M - MIPv6
          I1 - ISIS L1, I2 - ISIS L2, IA - ISIS interarea, IS - ISIS summary
          O - OSPF intra, OI - OSPF inter, OE1 - OSPF ext 1, OE2 - OSPF ext 2
          ON1 - OSPF NSSA ext 1, ON2 - OSPF NSSA ext 2
          D - EIGRP, EX - EIGRP external
S    2001:1:1:1::/64 [1/0]
     via 2001:2:1:1::2
C    2001:2:1:1::/64 [0/0]
     via ::, FastEthernet0/1
L    2001:2:1:1::1/128 [0/0]
     via ::, FastEthernet0/1
L    FF00::/8 [0/0]
     via ::, Null0
```

④ 在 V6_R1 上使用 show ipv6 router 命令验证 IPv6 路由：

```
V6_R1#sh ipv6 route
IPv6 Routing Table - 6 entries
Codes: C - Connected, L - Local, S - Static, R - RIP, B - BGP
       U - Per-user Static route, M - MIPv6
       I1 - ISIS L1, I2 - ISIS L2, IA - ISIS interarea, IS - ISIS summary
       O - OSPF intra, OI - OSPF inter, OE1 - OSPF ext 1, OE2 - OSPF ext 2
       ON1 - OSPF NSSA ext 1, ON2 - OSPF NSSA ext 2
       D - EIGRP, EX - EIGRP external
S    ::/0 [1/0]
     via 2001:2:1:1::1
C    2001:1:1:1::/64 [0/0]
     via ::, FastEthernet0/0
L    2001:1:1:1::1/128 [0/0]
     via ::, FastEthernet0/0
C    2001:2:1:1::/64 [0/0]
     via ::, FastEthernet0/1
L    2001:2:1:1::2/128 [0/0]
     via ::, FastEthernet0/1
L    FF00::/8 [0/0]
     via ::, Null0
```

注意查看路由表中加粗对象。

⑤ 在 BR 上 ping 2001:1:1:1::1/64 地址，抓包分析封装如图 7-7、图 7-8 所示。

```
At Device: BR
Source: BR
Destination: 2001:1:1:1::1
In Layers                          Out Layers
Layer7                             Layer7
Layer6                             Layer6
Layer5                             Layer5
Layer4                             Layer4
Layer3                             Layer 3: IPv6 Header Src. IP:
                                   2001:2:1:1::1, Dest. IP: 2001:1:1:1::1
                                   ICMPv6 Echo Message Type: 128
Layer2                             Layer 2: Ethernet II Header
                                   0090.212B.7E02 >> 0002.4A22.E802
Layer1                             Layer 1: Port(s): FastEthernet0/1
```

图 7-7　IOS 封装图

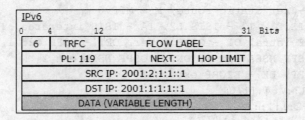

图 7-8 IPv6 封装图

(3) 配置 NAT-PT 实现 IPv4 和 IPv6 互通

① 在 BR 上配置：

```
interface fa0/0
  ipv6 nat
  exit
interface fa0/1
  ipv6 nat
  exit
ipv6 nat v4v6 source 172.30.0.1 2001:db1:0:1::2
ipv6 nat v6v4 source 2001:2:1:2::1 172.130.0.2
ipv6 nat prefix 2001:db1:0:1::/96
```

② 使用 show ipv6 nat translations 命令，验证 NAT-PT：

```
BR#sh ipv6 nat translations
Prot   IPv4 source          IPv6 source
       IPv4 destination     IPv6 destination
---    172.30.0.1           2001:DB1:0:1::2
---    172.130.0.2          2001:2:1:2::1
```

③ 在 CORE1 和 V6_R1 上使用 ping 命令，测试地址转换：

```
V6_R1 #ping 2001:db1:0:1::2
  Type escape sequence to abort.
  Sending 5, 100-byte ICMP Echos to 2001:db1:0:1::2, timeout is 2 seconds:
  !!!!!
  Success rate is 100 percent (5/5), round-trip min/avg/max = 60/73/120 ms
```

```
CORE1#ping 172.130.0.2
  Type escape sequence to abort.
  Sending 5, 100-byte ICMP Echos to 172.130.0.2, timeout is 2 seconds:
  !!!!!
  Success rate is 100 percent (5/5), round-trip min/avg/max = 64/73/92 ms
```

也可以在 BR 上使用 debug ipv6 nat 命令查看地址转换的过程。

## 工程化操作

① 保存 BR 当前配置为备份文件，使用 copy running-config flash 命令，目标文件为：BR_IPV4_BK。

② 针对 BR 进行 IPv6 部署。

创建脚本文件为 BR_IPv6_CFG.txt，内容如下：

```
##################      br-ipv6-cfg      ##################
configure terminal
ipv6 unicast-routing
interface fa0/0
ipv6 nat
exit
interface fa0/1
ipv6 address 2001:2:1:1::1/64
ipv6 nat
exit
ipv6 route 2001:1:1:1::/64 2001:2:1:1::2
ipv6 nat v4v6 source 172.30.0.1 2001:db1:0:1::2
ipv6 nat v6v4 source 2001:2:1:2::1 172.130.0.2
ipv6 nat prefix 2001:db1:0:1::/96 end
```

③ 针对 V6_R1 部署 IPv6。

创建脚本文件为 v6_R1_IPv6_CFG.txt，内容如下：

```
##################      v6_r1-ipv6-cfg      ##################
configure terminal
ipv6 unicast-routing
hostname V6_R1
ipv6 unicast-routing
interface fa0/1
ipv6 address 2001:2:1:1::2/64
exit
interface fa0/0
ipv6 address 2001:1:1:1::1/64
exit
ipv6 route ::/0 2001:2:1:1::1
```

④ 执行配置脚本。使用 CONSOLE 口或者远程连接登录设备，打开超级终端或者 SecureCRT 程序，进入特权模式，复制设备中的命令脚本（设备名_CFG.txt），在超级终端或者 SecureCRT 程序中，右击选择"粘贴"，执行脚本。

⑤ 验证配置。使用 show running-config 命令检查当前运行配置。

在 BR 上验证：

```
BR#show running-config
<省略部分>
ipv6 unicast-routing
!
interface FastEthernet0/0
 ip address 172.30.0.254 255.255.255.0
 duplex auto
 speed auto
 ipv6 nat
!
```

```
interface FastEthernet0/1
 no ip address
 duplex auto
 speed auto
 ipv6 address 2001:2:1:1::1/64
 ipv6 nat
!
interface Vlan1
 no ip address
!
ip classless
ip route 172.0.0.0 255.0.0.0 172.30.0.1
ip route 10.0.0.0 255.0.0.0 172.30.0.1
!
!
ipv6 nat v4v6 source 172.30.0.1 2001:DB1:0:1::2
ipv6 nat v6v4 source 2001:2:1:2::1 172.130.0.2
ipv6 nat prefix 2001:DB1:0:1::/96
<省略部分>
```

⑥ 确认无误后，在设备上保存配置，并保存设备配置脚本文件。

# 第 8 章 综合训练

校园网中网络升级与故障修复很常见，如新增部门、新增业务需求、技术升级，通常会涉及到链路、规划、安全、管理等环节，如果处理不好，将会给校园网造成不良影响。本章通过多个典型工程化项目中的问题的分析与解决，引导读者分析和解决校园网中的典型问题。

本章需要完成的项目有：

项目 16——物理架构与扩展；

项目 17——路由协议分析；

项目 18——新增应用服务；

项目 19——H3C 网络服务。

## 项目 16　物理架构与扩展

### 项目描述与设计

通过第 1 章（项目 1 和项目 2）的训练，我们对校园网（见图 8-1）的架构有了比较清晰的认识，由于受到 Packet Tracer 5.3 软件功能的限制，所有设备之间进行连接几乎都是用百兆电口或者千兆电口，核心层、汇聚层交换机都是用 3560-24 交换机，但是现实工程中两个层次设备选型肯定不同。

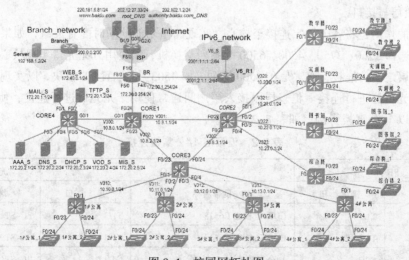

图 8-1　校园网拓扑图

本项目将对园区设备的实际连接进行进一步的分析,完成以下任务:
- 分析上行端口及链路选型与设计。
- 使用 Visio 绘制园区拓扑图。
- 对局部网络进行扩展。

1) 核心层与接口

核心层的主要功能就是保证校园网内数据高速转发,如果使用 CISCO 产品,可以选择 6500 或者 7600 系列交换机,同时使用万兆光口实现核心层之间、汇聚层与核心层之间的连接,通常是在 6500 系列(如 6513)上配置 TE 模块,以实现万兆连接并选择合适的光链路进行连接。以 CORE1 为例,如表 8-1 所示。

表 8-1 核心 1 连接

| FROM | | TO | | 光 纤 |
|---|---|---|---|---|
| 设备名称 | 接 口 | 设备名称 | 接 口 | 端口型号 |
| CORE 核心 C6513 | T1/1 | CORE2 核心 C6513 | T1/1 | SC-ST |
| | T1/2 | 数据中心汇聚 C4503 | T1/1 | SC-ST |
| | T1/3 | 1#公寓楼汇聚 C4503 | T1/1 | SC-ST |
| | T1/4 | 2#公寓楼汇聚 C4503 | T1/1 | SC-ST |
| | T2/1 | 3#公寓楼汇聚 C4503 | T1/1 | SC-ST |
| | T2/2 | 4#公寓楼汇聚 C4503 | T1/1 | SC-ST |
| | T2/3 | 教学楼汇聚 C4503 | T1/1 | SC-ST |
| | T2/4 | 实训楼汇聚 C4503 | T1/1 | SC-ST |
| | T3/1 | 图书馆汇聚 C4503 | T1/1 | SC-ST |
| | T3/2 | 综合楼汇聚 C453 | T1/1 | SC-ST |

使用双核心的目的,是为了实现单核故障不影响网络的可用性,通常使用 VRRP 实现容易,当然在实际项目设计中如果预算有限,可以在数据中心部分设计汇聚层到核心层的双上联。

2) 汇聚层与接口

汇聚层的作用是将各个楼宇连接至核心,实现路由功能的同时配置策略,如果使用 CISCO 产品,可以选择 3560、3750、4500、6500 等系列交换机,如使用万兆光口实现核心层之间、汇聚层与核心层之间的连接,通常是在 4500 系列(如 4503)上配置 TE 模块,以实现万兆连接并选择合适的光链路进行连接。汇聚层跟接入层之间通常使用千兆光口进行连接,所以配置 GE 模块即可。以 1#公寓楼为例,如表 8-2 所示。

表 8-2 汇聚层 1#公寓楼连接

| FROM | | TO | | 光 纤 |
|---|---|---|---|---|
| 设备名称 | 接 口 | 设备名称 | 接 口 | 端口型号 |
| 1#公寓楼汇聚 C4503 | T1/1 | CORE1 核心 C6513 | T1/3 | SC-ST |
| | T1/2 | CORE2 核心 C6513 | T1/3 | SC-ST |
| | G2/1 | 1#公寓楼接入 C2960 | G1/1 | SC-ST |

3）接入层与接口

接入层的主要作用是连接终端，主要考虑端口密度问题，使用光口模块上行连接。

## 项目任务 1：物理架构

① 登录 CISCO 官网，查阅 CISCO 交换机系列产品手册，分析校园网中的设备选型与模块、链路选择。

按照表 8-1 的格式，使用 Excel 制定全校网络设备连接表。

② 参考图 8-2（南京某高校校园网）所示结构，使用 Visio 2003 绘制校园网工程拓扑图。

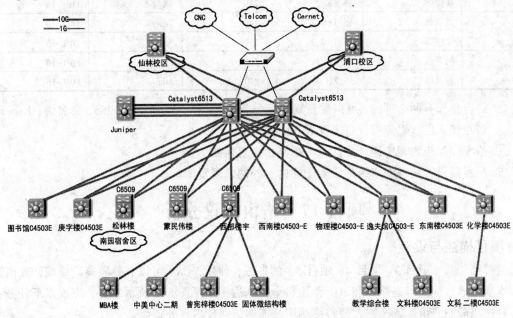

图 8-2　某高校校园网拓扑样图

## 项目任务 2：增加局部用户

某校申报国家级示范院校建设，下个月将有专家组进校进行验收与测评，学院安排专家组在综合楼 3 楼办公，考虑到不与现有用户冲突，考虑在该楼重新规划 VLAN 并分配独立网段地址，针对该需求请对网络进行改进。

① 分析原有 VLAN 规划，见表 8-3。

表 8-3　原有 VLAN 规划

| 楼　宇 | 部　门 | 设　备 | VLAN ID | VLAN NAME | 端　口 |
| --- | --- | --- | --- | --- | --- |
| 1#公寓 | A 座 | AS_1_GY_1 | 100 | 1_GY_1 | F0/1-16 |
|  | B 座 | AS_1_GY_2 | 110 | 1_GY_2 | F0/1-16 |
| 2#公寓 | A 座 | AS_2_GY_1 | 120 | 2_GY_1 | F0/1-16 |
|  | B 座 | AS_2_GY_2 | 130 | 2_GY_2 | F0/1-16 |
| 3#公寓 | A 座 | AS_3_GY_1 | 140 | 3_GY_1 | F0/1-16 |
|  | B 座 | AS_3_GY_2 | 150 | 3_GY_2 | F0/1-16 |

续表

| 楼 宇 | 部 门 | 设 备 | VLAN ID | VLAN NAME | 端 口 |
|---|---|---|---|---|---|
| 4#公寓 | A座 | AS_4_GY_1 | 160 | 4_GY_1 | F0/1-16 |
| | B座（1-3层） | AS_4_GY_2 | 170 | 4_GY_2_1 | F0/1-8 |
| | B座（4-6层） | | 180 | 4_GY_2_2 | F0/9-16 |
| 教学楼 | A楼 | AS_JX_1 | 200 | JX_1 | F0/1-16 |
| | B楼 | AS_JX_2 | 210 | JX_2 | F0/1-16 |
| 实训楼 | A楼 | AS_SX_1 | 220 | SX_1 | F0/1-16 |
| | B楼 | AS_SX_2 | 230 | SX_2 | F0/1-16 |
| 图书馆 | 教师部 | AS_TSG_1 | 240 | Teach | F0/1-16 |
| | 学生部 | AS_TSG_2 | 250 | Stu | F0/1-16 |
| 综合楼 | 财务部 | AS_ZH_1 | 260 | Finan | F0/1-16 |
| | 行政部 | AS_ZH_2 | 270 | Admin | F0/1-16 |

② 综合楼原有两个部门，可以重新划分一个新的 VLAN，如编号为：265，命名为：Expert。
③ 请在接入层、汇聚层设备上配置 VLAN。
④ 给 VLAN 规划相应地址（注意考虑路由汇聚问题）。
⑤ 配置路由，使得专家组网段能访问内、外网常规资源。

## 项目17　路由协议分析

### 项目描述与设计

通过第 2 章（项目 3、项目 4、项目 5）的训练，使用 VLAN 可以有效隔离广播域，方便管理，本项目通过配置静态路由和 RIP 路由将全网联通，按照图 8-3 所示结构，完成以下任务：

- 使用静态路由配置全网，使得终端用户能够访问校内、外资源。
- 使用动态路由协议（RIP）配置网络。
- 分析 RIP 协议的优缺点。

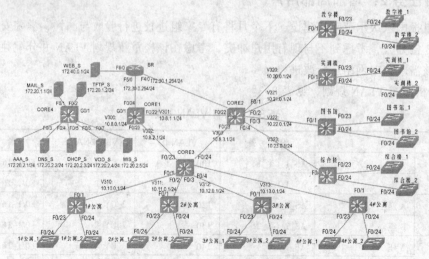

图 8-3　PACKET TRACERT 中拓扑图

项目任务 1：使用静态路由配置全网

1）网络编址

IP 地址规划表如表 8-4 所示。

表 8-4　IP 地址规划表

| 设备名称 | 端口 | IP 地址 | 掩码 | 设备名称 | 端口 | IP 地址 | 掩码 |
| --- | --- | --- | --- | --- | --- | --- | --- |
| DS_1_GY | V310 | 10.10.0.254 | 255.255.255.0 | DS_ZH | V323 | 10.23.0.254 | 255.255.255.0 |
| | V100 | 172.16.0.254 | 255.255.255.0 | | V260 | 172.17.24.254 | 255.255.255.0 |
| | V110 | 172.16.4.254 | 255.255.255.0 | | V270 | 172.17.28.254 | 255.255.255.0 |
| DS_2_GY | V311 | 10.11.0.254 | 255.255.255.0 | CORE1 | F0/24 | 172.30.0.1 | 255.255.255.0 |
| | V120 | 172.16.8.254 | 255.255.255.0 | | V300 | 10.8.0.1 | 255.255.255.0 |
| | V130 | 172.16.12.254 | 255.255.255.0 | | V301 | 10.8.1.1 | 255.255.255.0 |
| DS_3_GY | V312 | 10.12.0.254 | 255.255.255.0 | CORE2 | V302 | 10.8.2.1 | 255.255.255.0 |
| | V140 | 172.16.16.254 | 255.255.255.0 | | F0/24 | 172.30.1.1 | 255.255.255.0 |
| | V150 | 172.16.20.254 | 255.255.255.0 | | V301 | 10.8.1.254 | 255.255.255.0 |
| DS_4_GY | V313 | 10.13.0.254 | 255.255.255.0 | | V303 | 10.8.3.1 | 255.255.255.0 |
| | V160 | 172.16.24.254 | 255.255.255.0 | | V320 | 10.20.0.1 | 255.255.255.0 |
| | V170 | 172.16.28.254 | 255.255.255.0 | | V321 | 10.21.0.1 | 255.255.255.0 |
| | V180 | 172.16.32.254 | 255.255.255.0 | | V322 | 10.22.0.1 | 255.255.255.0 |
| DS_JX | V320 | 10.20.0.254 | 255.255.255.0 | CORE3 | V323 | 10.23.0.1 | 255.255.255.0 |
| | V200 | 172.17.0.254 | 255.255.255.0 | | V302 | 10.8.2.254 | 255.255.255.0 |
| | V210 | 172.17.4.254 | 255.255.255.0 | | V303 | 10.8.3.254 | 255.255.255.0 |
| DS_SX | V321 | 10.21.0.254 | 255.255.255.0 | | V310 | 10.10.0.1 | 255.255.255.0 |
| | V220 | 172.17.8.254 | 255.255.255.0 | | V311 | 10.11.0.1 | 255.255.255.0 |
| | V230 | 172.17.12.254 | 255.255.255.0 | | V312 | 10.12.0.1 | 255.255.255.0 |
| DS_TSG | V323 | 10.23.0.254 | 255.255.255.0 | CORE4 | V313 | 10.13.0.1 | 255.255.255.0 |
| | V240 | 172.17.16.254 | 255.255.255.0 | | V500 | 172.20.1.254 | 255.255.255.0 |
| | V250 | 172.17.20.254 | 255.255.255.0 | | V501 | 172.20.2.254 | 255.255.255.0 |

2）配置静态|默认路由

① 在每个三层设备上使用 ip route 命令配置非直连路由，并使用 show ip route 命令进行验证。

② 配置默认路由，保证终端用户能访问外网资源。详细步骤可参考第 1 章项目 2 的配置脚本。

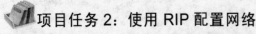 项目任务 2：使用 RIP 配置网络

静态路由在大型网络中的配置工作量繁重，不利于检查与排错，本任务中使用 RIP，并使用 Version 2，来部署校园网体现 RIPv2 的增强特性。

1）基本配置

在每个设备上配置：

```
设备名（config）#router rip
设备名（config-rip）#version 2
设备名（config-rip）#no auto-summary
设备名（config-rip）#network 直连网络主类地址
```

2）发布默认路由

在出口设备上配置（如 CORE1）：

```
CORE1（config-rip）#default-information originate
```

3）分析 RIP 收敛

① 当 RIP 路由收敛后，在 CORE1 上将连接外网端口，使用 shoudown 命令关闭。

② 在教学楼汇聚交换机，CORE4、1#公寓楼汇聚交换机上使用 debug ip rip 命令查看 RIP 运行过程。

③ 在教学楼汇聚交换机，CORE4、1#公寓楼汇聚交换机上使用 show iproute 命令查看路由表动态过程。

④ 当 RIP 路由重新收敛后，在 CORE1 上将连接外网端口，使用 no shoudown 命令开启。

⑤ 在教学楼汇聚交换机，CORE4、1#公寓楼汇聚交换机上再次查看 RIP 运行过程。

⑥ 在教学楼汇聚交换机，CORE4、1#公寓楼汇聚交换机上再次查看路由表动态过程。

总结：RIP 协议链路切换时，路由收敛太慢，协议存在明显缺陷。

4）使用 OSPF 路由协议

详细步骤可参考第 3 章项目 6 的内容。

# 项目 18　新增应用服务

## 项目描述与设计

随着高校数字化建设的发展，一卡通系统已成为很多高校的典型应用，考虑到数据的安全性和保密性，很多学校采用专网方式进行（物理上跟校园网隔开），这种方式相对而言建设成本太高，因此在校园网中建立虚拟专网成为一卡通的一种解决方案。

一卡通系统遍及每个楼宇，出于安全角度，每个楼宇需要划出一个独立的网段，保证这些网段能通信畅通，而跟其他网段逻辑上隔离。

典型的解决方法有两种：

第一，将专网使用静态路由方式进行配置，不参与校园网 OSPF 的集成。

第二，配置访问控制列表，限制网络之间的访问。

## 项目任务 1：规划各个楼宇网段

若原有设计中预留了部分扩展空间（校园网地址规划设计应该考虑到以后的升级），可以在每栋楼中增加相应的 VLAN，如在 1#公寓楼增加一卡通网段，如表 8-5 所示。

表 8-5　1#公寓楼增加一卡通网段

| 设备名称 | 端口 | IP地址 | 掩码 |
|---|---|---|---|
| DS_1_GY | V310 | 10.10.0.254 | 255.255.255.0 |
|  | V100 | 172.16.0.254 | 255.255.255.0 |
|  | V101 | 172.16.1.254 | 255.255.255.0 |
|  | V110 | 172.16.4.254 | 255.255.255.0 |

请完善下表 8-6，以增加网段。

表 8-6　全网增加一卡通网段

| 设备名称 | 端口 | IP地址 | 掩码 | 设备名称 | 端口 | IP地址 | 掩码 |
|---|---|---|---|---|---|---|---|
| DS_1_GY | V310 | 10.10.0.254 | 255.255.255.0 | DS_JX | V320 | 10.20.0.254 | 255.255.255.0 |
|  | V100 | 172.16.0.254 | 255.255.255.0 |  | V200 | 172.17.0.254 | 255.255.255.0 |
|  | V101 | 172.16.1.254 | 255.255.255.0 |  |  |  |  |
|  | V110 | 172.16.4.254 | 255.255.255.0 |  | V210 | 172.17.4.254 | 255.255.255.0 |
| DS_2_GY | V311 | 10.11.0.254 | 255.255.255.0 | DS_SX | V321 | 10.21.0.254 | 255.255.255.0 |
|  | V120 | 172.16.8.254 | 255.255.255.0 |  | V220 | 172.17.8.254 | 255.255.255.0 |
|  | V130 | 172.16.12.254 | 255.255.255.0 |  | V230 | 172.17.12.254 | 255.255.255.0 |
| DS_3_GY | V312 | 10.12.0.254 | 255.255.255.0 | DS_TSG | V323 | 10.23.0.254 | 255.255.255.0 |
|  | V140 | 172.16.16.254 | 255.255.255.0 |  | V240 | 172.17.16.254 | 255.255.255.0 |
|  | V150 | 172.16.20.254 | 255.255.255.0 |  | V250 | 172.17.20.254 | 255.255.255.0 |
| DS_4_GY | V313 | 10.13.0.254 | 255.255.255.0 | DS_ZH | V323 | 10.23.0.254 | 255.255.255.0 |
|  | V160 | 172.16.24.254 | 255.255.255.0 |  | V260 | 172.17.24.254 | 255.255.255.0 |
|  | V170 | 172.16.28.254 | 255.255.255.0 |  |  |  |  |
|  |  |  |  |  | V270 | 172.17.28.254 | 255.255.255.0 |
|  | V180 | 172.16.32.254 | 255.255.255.0 |  |  |  |  |

## 项目任务 2：使用静态路由联通

在全网运行 OSPF 路由的前提下，使用静态路由方式联通一卡通网段。

 注意

不可让校园其他网络参与进来！

## 项目任务 3：使用 ACL 进行控制

让所有一卡通网段参与 OSPF 路由，使用 ACL 进行访问控制，以实现一卡通网段与校园网常规网段在逻辑上隔离。

# 项目 19  H3C 解决方案

## 项目描述与设计

校园网的建设通常经过多期才能逐步完善，在各期项目中几乎很少有学校全都用同一厂商（如 CISCO）的产品，因此在校园网中会出现各个厂商的产品，从支持民族产业和性价比角度来考虑，国产设备更具有优势，如华为、H3C、锐捷等。

本项目主要考虑以下内容：

- 全网使用 H3C 产品，要求能阅读 H3C 产品手册，进行选型。
- 基于 H3C 操作系统配置，编写相关报告。

## 项目任务 1：设备选型

① 登录 H3C 官网（http://www.h3c.com.cn），查看产品中的园区交换机产品系列，比较二层、三层、核心交换机的技术指标，可以跟 CISCO 产品做详细比较。

② 了解 H3C 的产品特性协议，与 CISCO 产品进行比较。

③ 参考官网"解决方案"中的"教育"部分，进行设备初步选型。

④ 官网如没有要查阅的资料，可以拨打 400 电话咨询。

## 项目任务 2：方案设计

① 根据设备选型参考项目 17 的要求，绘制拓扑图。

② 使用 Excel 编写表 8-7、表 8-8 和表 8-9。

表 8-7  设备配置表

| 序号 | 设备名 | 数量 | 型号 | 序列号 | 用途 | 备注 |
|---|---|---|---|---|---|---|
| 1 | | | | | | |
| 2 | | | | | | |
| 3 | | | | | | |
| 4 | N_2_C_S3952 | 1 | Quidway S3952P-EI | | 北 2c 接入 | 样例 |

表 8-8  设备互连表

| 序号 | 本端设备名称 | 本端接口 | 对端设备名称 | 对端接口 | 链路类型 | IP/trunk | 备注 |
|---|---|---|---|---|---|---|---|
| | | | | | | | |
| | | | | | | | |
| | | | | | | | |
| 4 | N_1_B_s3952 | G1/1/4 | N_1_C_S3952 | G1/1/3 | | trunk | 样例 |

表 8-9  设备管理表

| 序号 | 设备名称 | 管理 IP | 管理账号 | 管理密码 | 登录方式 | 备注 |
|---|---|---|---|---|---|---|
| | | | | | | |
| | | | | | | | 
| 4 | N_2_C_S3952 | 192.168.104.21 | ****** | ***** | telnet | 样例 |

# 项目1  建设校园网基础训练图

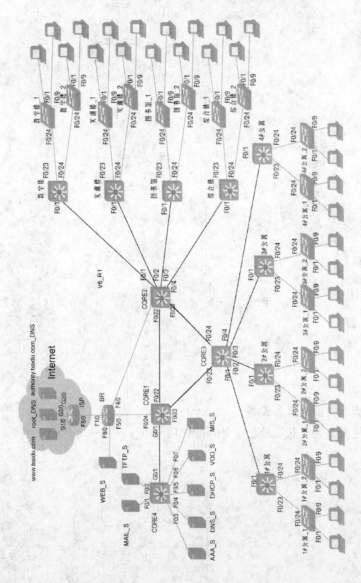

训练内容：

在 Packet tracer 5.3 中设计校园网布局，并将各楼宇、设备间设备进行连接。

# 项目 2　配置校园网训练图

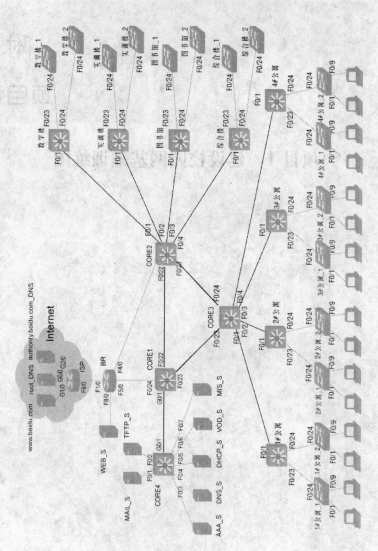

训练内容：
① 按照脚本文件配置各设备；
② 检查终端是否能获取地址、是否能访问 web_s。

# 项目 3  部署接入 VLAN 训练图

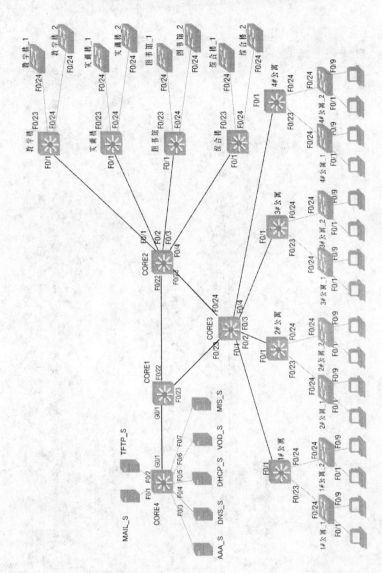

训练内容：
① 在接入层交换机设计规划 VLAN；
② 在接入层交换机配置并验证 VLAN。

# 项目 4　部署汇聚 VLAN 训练图

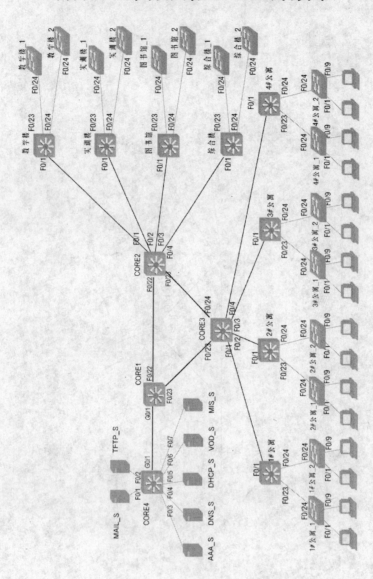

训练内容：
① 在汇聚层交换机设计规划 VLAN；
② 在汇聚层交换机配置并实现 VLAN 互连。

## 项目 5  部署管理 VLAN 训练图

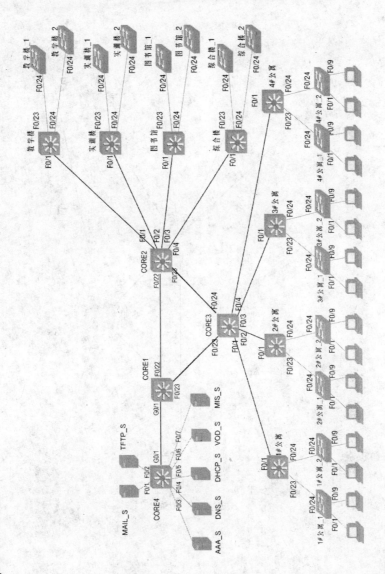

训练内容：
① 统一规划管理 VLAN 以实现二层管理；
② 在核心、汇聚、接入各层部署管理 VLAN。

# 项目6　优化 OSPF 路由训练图

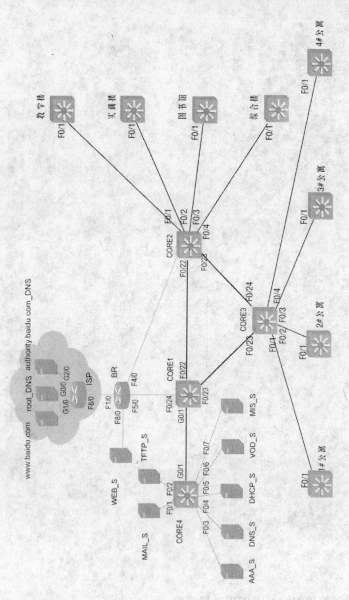

训练内容：
① 使用 OSPF 部署校园网络由；
② 利用 stub、nssa 等特性优化路由。

# 项目 7　访问按控制管理训练图

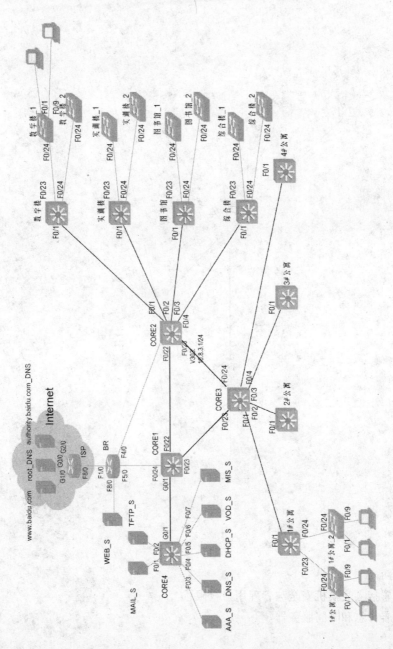

训练内容：
① 使用 ACL 优化网络内部访问；
② 使用防病毒 ACL 保护网络安全。

# 项目 8  内网安全训练图

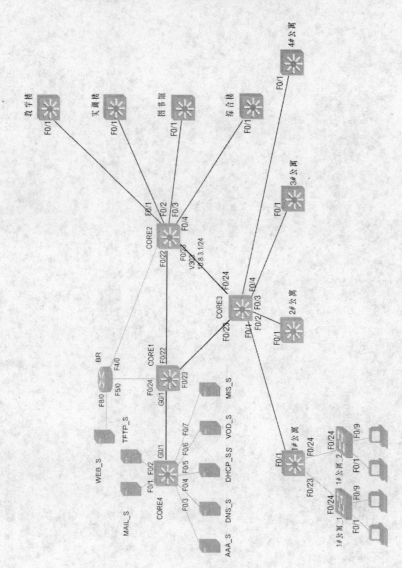

训练内容：
① 配置接入交换机安全与管理用户权限；
② 部署日志和时间服务器。

# 项目9  部署 DHCP 系统训练图

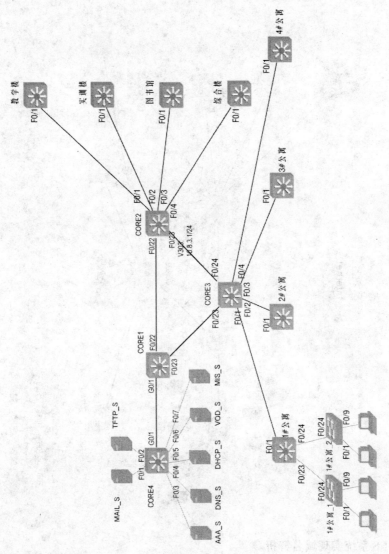

训练内容:
① 部署 DHCP 服务器;
② 配置 DHCP 安全与防护。

# 项目 10　部署 DNS 系统训练图

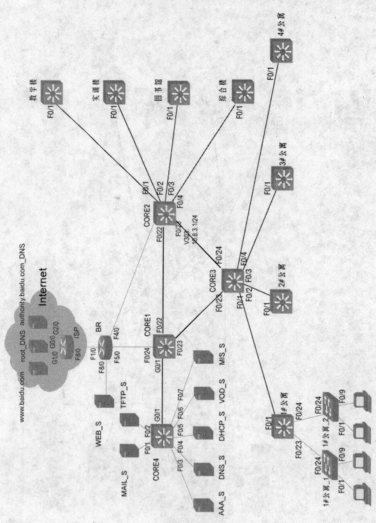

训练内容：
　　部署 DNS 系统实现域名解析。

# 项目 11　部署 VPN 训练图

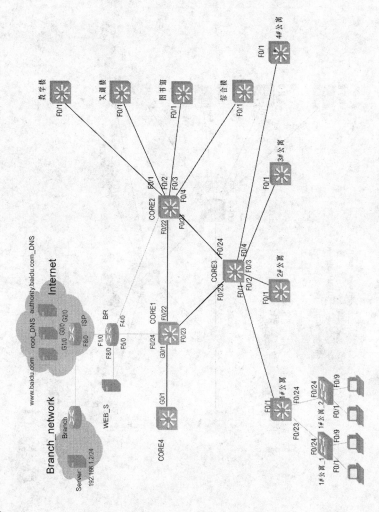

训练内容：
① 部署 site to site IPSec VPN 以实现分部与总部通信；
② 部署 remote VPN 以实现外网用户访问内网资源。

# 项目12  部署防火墙训练图

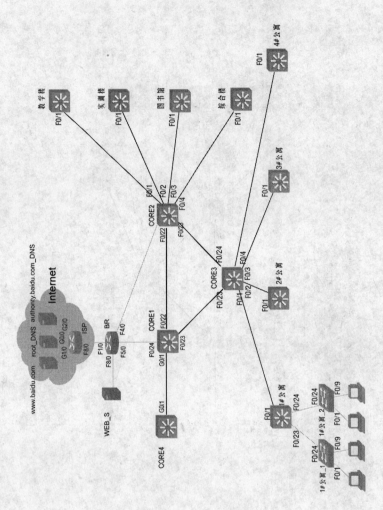

训练内容：

使用 CBAC 技术部署防火墙。

# 项目 13　使用管理工具训练图

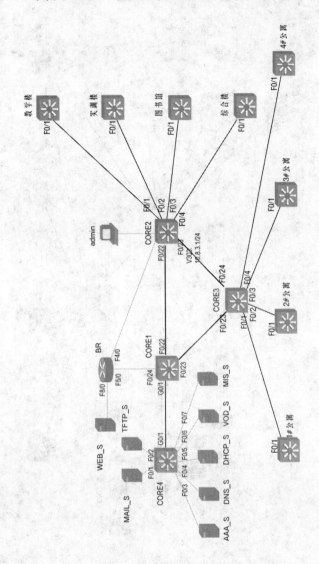

训练内容：
① 使用 telnet、ssh、snmp 管理网络；
② 使用 tftp 服务器备份恢复配置。

# 项目14  AAA部署训练图

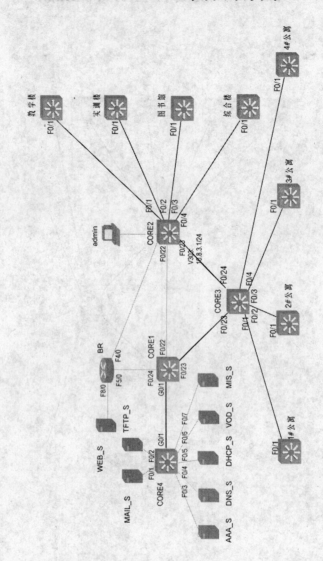

训练内容：
使用AAA服务器统一对设备管理账户统一管理。

# 项目 15　IPv6 实验网训练图

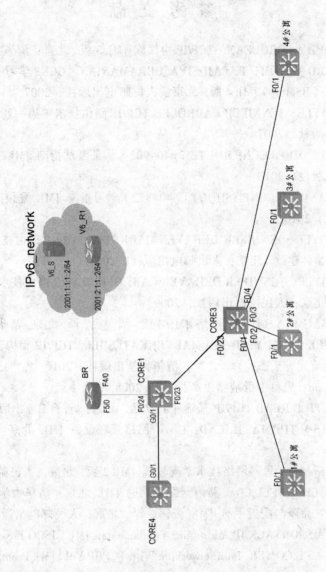

训练内容：
① 配置设备 IPv6 地址；
② 配置 IPv6 路由，并实现 IPv6 与 IPv4 过渡。

# 参 考 文 献

[1] [美] BEHROUZ A.FOROUZAN. TCP/IP 协议族 [M]. 3 版. 北京：清华大学出版社，2009.

[2] [美] RICHARD FROOM，BALAJI SIVASUBRAMANIAN. CCNP 学习指南：组建 Cisco 多层交换网络（BSMSN）[M]. 4 版. 北京：人民邮电出版社，2007.

[3] [美]JEFF DOYLE，JENNIFER CARROLL. TCP/IP 路由技术（第一卷）[M]. 2 版. 北京：人民邮电出版社，2007.

[4] [美]WENDELL ODOM.CCNP ROUTE（640-902）认证考试指南 [M]. 王兆文，译. 北京：人民邮电出版社，2010.

[5] [美]AMIR RANJBAR.CCNP TSHOOT（640-832）学习指南 [M]. 夏俊杰，译. 北京：人民邮电出版社，2010.

[6] [美]JEFF DOYLE，JENNIFER DEHAVEN CARROLL. TCP/IP 路由技术（第一卷）[M]. 2 版. 葛建立，吴建章. 北京：人民邮电出版社，2007.

[7] [美]JEFF DOYLE，JENNIFER DEHAVEN CARROLL. TCP/IP 路由技术（第二卷）[M]. 夏俊杰，译. 北京：人民邮电出版社，2009.

[8] [美]TANENBAUM，A.S.计算机网络 [M]. 4 版. 潘爱民，译. 北京：清华大学出版社，2004.

[9] [印度]SAMEER SETH，M.AJAYKUMARVENKATESULU. TCP/IP 架构、设计及应用（Linux 版）[M]. 黄清元，于杰，译. 北京：清华大学出版社，2010.

[10] CISCO System. 思科中低端路由产品手册. 2008.

[11] [美]KENNETH D.READ.TCP/IP 基础 [M]. 7 版. 北京：电子工业出版社，2003.

[12] [美]THOMAS M THOMAS II.ICND：Cisco 网络互联设备 [M]. 北京：机械工业出版社，2001.

[13] [美]CISCO System. 思科网络技术学院教程 [M]. 2 版. 北京：人民邮电出版社，2002.

[14] [美]LAURA CHAPPELL.Cisco 路由器配置导论 [M]. 北京：清华大学出版社，1999.

[15] CISCO OSPF 命令与配置手册 [M]. 张伟，译. 北京：人民邮电出版社，2003.

[16] [美]MARK A.SPORTACK.IP Addressing Fundamentals [M]. CISCO Press，2002.

[17] [美]DOUGLAS E.COMER. Internetworking With TCP/IP Vol I [M]. Prentice Hall Press，2002.

[18] [美]FARAZ SHAMIM.Troubleshooting IP Routing Protocols [M]. Cisco Press，2001.

[19] [美]David Hucaby.Cisco Field Manual：Catalyst Switch Configurat-ion [M]. Cisco Press，2002.

[20] 思科官网：http://www.cisco.com.

[21] CISCO Packet Tracer 5.3 Tutorials.

[22] 思科网院 CCNA Discovery 4.0 课程.

[23] 思科网院 CCNA Expore 4.0 课程.

[24] 思科网院 CCNA Security 1.0 课程.

[25] Cisco Product Quick Reference Guide August 2010 edition.

[26] Castalyst 6500 Series Switch Command Reference Release 7.6.

[27] Catalyst 2950 and Catalyst 2955 Switch Command Reference, 12.1(22)EA11 and Later.

[28] Catalyst 2960 Switch Command Reference, 12.2(25)SED.

[29] Catalyst 3560 Switch Command Reference, Rel. 12.2(25)SEA.

[30] Catalyst 3750 Switch Command Reference, 12.2(25)SEA.

[31] Catalyst 3750 Command Reference, Cisco IOS Release 12.2(58)SE.

[32] Catalyst 6500 Series Switch Content Switching Module Command Reference Software Relese 4.1(2).

[33] Configuration: Basic Software Configuration Using the Cisco IOS Command-Line Interface.

[34] Cisco ASA 5500 Series Command Reference, 8.3.

[35] H3C S5500-EI 系列以太网交换机 命令手册, Release 2102.

[36] H3C S3100 系列以太网交换机 操作手册, Release 21XX 系列（V1.07）.

笔记栏